PRINCIPLES OF COMBINATORICS

This is Volume 72 in
MATHEMATICS IN SCIENCE AND ENGINEERING
A series of monographs and textbooks
Edited by RICHARD BELLMAN, *University of Southern California*

A complete list of the books in this series appears at the end of this volume.

PRINCIPLES OF COMBINATORICS

C. BERGE

CHAIRE DE CALCUL DES PROBABILITÉS
ET PHYSIQUE MATHÉMATIQUE
UNIVERSITÉ DE PARIS
PARIS, FRANCE

 1971

ACADEMIC PRESS   *New York and London*

First published in the French language under the title *Principes de Combinatoire*.
©1968 by Dunod, Paris, France.

ACADEMIC PRESS, INC.
111 Fifth Avenue, New York, New York 10003

United Kingdom Edition published by
ACADEMIC PRESS, INC. (LONDON) LTD.
Berkeley Square House, London W1X 6BA

LIBRARY OF CONGRESS CATALOG CARD NUMBER: 76-127679
AMS 1970 Subject Classifications 05-02, 05-01, 05A15, 05A17, 05C30

PRINTED IN THE UNITED STATES OF AMERICA

CONTENTS

v

3. Inversion Formulas and Their Applications

4. Permutation Groups

5. Pólya's Theorem

FOREWORD

Most mathematicians of this day, confronted with an argument requiring combinatorial thinking, react with one of two stock phrases: (a) "This is a purely combinatorial argument," (b) "This is a difficult combinatorial argument." Hypnotic repetition of either of these slogans is likely to have the same balming effect on the speaker: freed from all scruples, he will pass the buck and unload the work onto someone else's shoulders.

While the end result of this oft-repeated vignette is an overwhelming variety of problems for specialists in the art, the impression grows that among mathematicians, especially "pure" mathematicians, total ignorance of combinatorial theory is as proudly flaunted as—in bygone days—an aristocrat's ignorance of his country's vernacular.

It is tempting to react to this rejection, which in the past has succeeded in finessing combinatorialists into the mathematical proletariat, by a ringing *ça ira*. This might well take the form of a concerted attack on one of the currently fashionable branches of mathematics. The barrage of definitions and the superstructure of grammatical gamesmanship removed, little more than a few puny combinatorial facts would be left, which would then be dealt an embarrassingly easy *coup de grace* by the application of standard combinatorial techniques.

Fortunately, this course will not be followed, for a sound reason: combinatorialists have better fish to fry. In the last ten years—shortly after the appearance of Claude Berge's first book—the field has increased geometrically in output, depth, and importance. The shoe will soon be on the other foot, if it isn't already.

Two Frenchmen have played a major role in the renaissance of combinatorics: Berge and Schützenberger. Berge has been the more prolific writer, and his books have carried the word farther and more effectively than anyone anywhere. I recall the pleasure of reading the disparate examples in his first book, which made it impossible to forget the material. Soon after that reading, I would be one of many who unknotted themselves from the tentacles of the Continuum and joined the then Rebel Army of the Discrete.

What a renewed pleasure is it to again read Berge in the present introductory book! The material is roughly complementary to that of his previous books, and deals largely with enumeration. The choice of topics is perfectly balanced, the presentation impeccably elegant, and the material can be followed by anyone with an interest in the subject and with only a little algebra as background. Some topics are here collected in book form for the first time, notably the beautiful Robinson–Shensted theorem, the Eden–Schützenberger decomposition theorem, and various facts connecting Young diagrams, trees, and the symmetric group.

One regrets that, in order to minimize the book's length, a full presentation of the theory of representations of the symmetric group along purely combinatorial lines could not be given; but then, this would probably take a book by itself. What is given is enough to whet the reader's appetite and lead him to other sources—an ample bibliography is given—for fuller information.

Perhaps the main virtue of this book is its inspirational, uplifting quality. The old Satz–Beweis style of mathematical exposition, which turned the reader into a code cracker, is irretrievably going—algebraists take heed—and is being replaced by a discursive, example-rich flow reminiscent of the classics of yesteryear: Salmon, Weber, Bertini . . . .

Berge is a master of this expository *nouvelle vague*. I am tempted to suggest that the title of this book be changed to "Seduction into Combinatorics."

Gian-Carlo Rota

PRINCIPLES OF COMBINATORICS

WHAT IS COMBINATORICS?

Nowadays, combinatorial analysis (or "combinatorics") is the focus of much attention; yet, nowhere in the literature does there seem to be a satisfactory definition of this science and its many ramifications.

In fact, mathematicians feel instinctively that certain problems are of a "combinatorial nature," and that the methods for solving them deserve to be studied systematically. In any case, this is what Pólya said in the introduction to the first volume of the *Journal of Combinatorial Theory*.*

We wish to offer here a definition of combinatorics, which depends on the very precise concept of "configuration."

A *configuration* arises every time objects are distributed according to certain predetermined constraints. Cramming miscellaneous packets into a drawer is an example of a configuration.

Consider, for example, two mutually orthogonal Latin squares:

Aa	Gh	Fi	Ej	Jb	Id	Hf	Bc	Ce	Dg
Hg	Bb	Ah	Gi	Fj	Jc	Ie	Cd	Df	Ea
If	Ha	Cc	Bh	Ai	Gj	Jd	De	Eg	Fb
Je	Ig	Hb	Dd	Ch	Bi	Aj	Ef	Fd	Gc
Bj	Jf	Ia	Ac	Ee	Dh	Ci	Fg	Gb	Ad
Di	Cj	Jg	Ib	Hd	Ff	Eh	Ga	Ac	Be
Fh	Ei	Dj	Ja	Ic	He	Gg	Ab	Bd	Cf
Cb	Dc	Ed	Fe	Gf	Ag	Ba	Hh	Ii	Jj
Ec	Fd	Ge	Af	Bg	Ca	Db	Ij	Jh	Hi
Gd	Ae	Bf	Cg	Da	Eb	Fc	Ji	Hj	Ih

* Published by Academic Press, New York, this journal was initiated in 1966.

In this case, the objects are the elements of the Cartesian product

$$\{A, B, C, D, E, F, G, H, I, J\} \times \{a, b, c, d, e, f, g, h, i, j\},$$

which are mapped into the 100 "boxes" of the 10 × 10 square. The constraints are as follows:

(1)   the mapping is a bijection, i.e., each box contains one and only one object, and every object appears once in the square;

(2)   no letter, large or small, occurs more than once in the same row or column.

This square is one of the most remarkable configurations in the history of combinatorics, since Euler conjectured that it did not exist. Euler's conjecture was refuted only in 1960 by Bose, Shrikhande, and Parker [14], who proved the existence of such squares for all orders other than six.

The concept of configuration can be made mathematically precise by defining it as a *mapping of a set of objects into a finite abstract set with a given structure*; for example, a permutation of n objects is a "bijection of the set of n objects into the ordered set 1, 2, ..., n." Nevertheless, one is only interested in mappings *satisfying certain constraints*.

Just as arithmetic deals with integers (with the standard operations), algebra deals with operations in general, analysis deals with functions, geometry deals with rigid shapes, and topology deals with continuity, so does combinatorics deal with configurations. Combinatorics counts, enumerates,* examines, and investigates the existence of configurations with certain specified properties. With combinatorics, one looks for their intrinsic properties, and studies transformations of one configuration into another, as well as "subconfigurations" of a given configuration.

The preoccupations of combinatorics are exactly the same as those of the other branches of modern mathematics. Nevertheless, it is surprising that this particular discipline has developed on the edge of, or away from, the mainstream of modern mathematics. The elementary theorems of the subject have been forgotten and rediscovered several times. The polycephalic mathematician Nicolas Bourbaki hardly mentions Pólya's theorem. In his some 20 volumes already published, there are very few general combinatorial theorems, although many

* Refer to the introduction, fifth aspect.

fundamental combinatorial formulas are distilled throughout the text. It is important to rectify this.

FIRST ASPECT: STUDY OF A KNOWN CONFIGURATION

The first phase in the development of combinatorial mathematics consists of studying the intrinsic properties of a given, or easily constructed, configuration. From the beginning of our era, the students of the divinatory system of geomancy [1] have studied aleatory configurations. In a letter purported to have been addressed by Archimedes to Eratosthenes of Cyrene, it is proposed, subject to certain conditions, to "compute the number of cattle of the Sun." Part of the statement of the problem is*:

When the white bulls mingled their number with the black, they stood firm, equal in depth and breadth, and the plains of Thrinacia, stretching far in all ways, were filled with their multitude. Again, when the yellow and the dappled bulls were gathered into one herd they stood in such a manner that their number beginning from one, grew slowly greater till it completed a triangular figure, there being no bulls of other colours in their midst nor any of them lacking [2].

This problem, one of the rare allusions in antiquity to combinatorics, depends, in fact, on arithmetic considerations, in particular, the polygonal numbers of Pythagorus, Nicomachus, and Diophantus. We have to wait for Euler before a school of authentic combinatorial mathematics appears.

SECOND ASPECT: INVESTIGATION OF AN UNKNOWN CONFIGURATION

Investigating the existence or nonexistence of a configuration with certain specified properties is another aspect of combinatorics. This is the situation in the famous problem of the nine bridges of the

* It is very remarkable that this problem leads to Fermat's equation:

$$y^2 - 410,286,423,278,424x^2 = 1.$$

The smallest solution for the total strength of the herd would be of order $7766 \times 10^{206,541}$.

town of Königsberg (now called Kaliningrad), or in that of the 36 officers of Euler, or, in a more abstract setting, in the construction of finite geometries. It is rather disconcerting to discover this type of investigation in the *Yi-King* [3], the divinatory book used in China by the lesser Taoists, and one of the oldest (about 2200 B.C.) texts still extant. This sacred work describes two configurations: The "Grand Plan" (Lo-Shu) and the "River Map." The "Grand Plan" which, legend claims, was decorated upon the back of a divine tortoise that emerged from the River Lo, is illustrated in Fig. 1; by substituting integers for the various sets of marks, we obtain the famous magic square "Saturn"

$$4 \quad 9 \quad 2$$
$$3 \quad 5 \quad 7$$
$$8 \quad 1 \quad 6$$

This configuration is remarkable since the sum of the elements in any row, column, or diagonal is always the same, that is, 15.

The "River Map," which, again according to legend, was found decorated upon the back of a divine tortoise that emerged from a river (this time the River Ho), is reproduced in Fig. 2; substituting integers we obtain the following configuration:

$$7$$
$$2$$
$$10 \quad\quad 10$$
$$8 \quad 3 \quad\quad 5 \quad\quad 4 \quad 9$$
$$10 \quad\quad 10$$
$$1$$
$$6$$

The symmetry, relative to the center, of the sums of adjacent digits in this map is best seen by inspection, for example,

$5 + 3 = 8$; $5 + 1 = 6$; etc. $3 + 10 + 2 = 8 + 7$; $3 + 10 + 1 = 8 + 6$; etc.

Knowing the difficulty of constructing such configurations, one could not possibly say that combinatorics did not exist in Chinese antiquity.

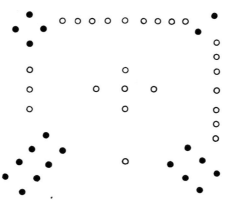

FIG. 1. The Grand Plan.

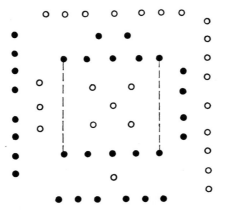

FIG. 2. The River Map.

THIRD ASPECT: COUNTING CONFIGURATIONS

For certain easily obtainable configurations, such as the combinations of n objects taken p at a time, it is natural to ask how many of them there are; hence, a new development in the evolution of combinatorics is concerned with obtaining exact formulas for the number of configurations satisfying certain specified properties.

Of course, combinatorics has developed in this direction as a result of the powerful influences of statistics and the calculus of probabilities (by the very definition of "probability"). Indeed, for a long time, most mathematicians identified combinatorics with "counting." For Riordan [6], its main function is "finding the number of ways there are of doing some well-defined operation."

If we limit ourselves to this viewpoint, then almost everywhere we can see the abortive beginnings of the subject, the majority of its formulas having been discovered and rediscovered many times over. The foremost example of such beginnings are the binomial coefficients, which were known to the twelfth-century school of the Indian arithmetician Bhaskra; nevertheless, the western world remained in ignorance of them until Pascal and Fermat rediscovered them as a byproduct of studying games of chance. Recently, it was discovered that the recurrence method for obtaining binomial coefficients and "Pascal's triangle" had been taught in 1265 by a Persian philosopher, Nasir-Ad-Din [7]. We also know that Cardan, in about 1560, proved that the number of subsets of a set of n elements is 2^n.

In 1666, Leibniz, at the age of 20, published the first treatise in combinatorics: "Dissertatio de Arte Combinatoria." He explained in the preface how he foresaw a new discipline having ramifications in logic, history, and even in morals, metaphysics, and "the whole range of the Sciences"!

As the configurations became more complex, combinatorialists, starting principally with Euler, became progressively interested in actual counting techniques; most probabilists, statisticians and engineers interested in counting have reaped the benefits. The most celebrated discoveries along these lines are the generating functions, which were discovered by Laplace, but had already been used implicitly by Euler; and the theorem attributed to Pólya, but apparently anticipated [4] by Redfield [5]. At any rate, this is the claim made by Blanche Descartes in a poem that we shall quote in full:

"ENUMERATIONAL"

Pólya had a theorem
(Which Redfield proved of old).
What secrets sought by graphmen
Whereby that theorem told!

So Pólya counted finite trees
(As Redfield did before).
"Their number is exactly such,
And not a seedling more."

Harary counted finite graphs
(Like Redfield, long ago),
And pointed out how very much
To Pólya's work we owe.

And Read piled graph on graph on graph
(Which is what Redfield did).
So numbering the graphic world
That nothing could be hid.

Then hail, Harary, Pólya, Read,
Who taught us graphic lore,
And spare a thought for Redfield too,
Who went too long before.

—BLANCHE DESCARTES

FOURTH ASPECT: APPROXIMATE COUNTING OF CONFIGURATIONS

While combinatorics was concerned only with "counting," the science remained a collection of scattered results, although this essentially workman-like viewpoint produced numerous very beautiful and frequently surprising arithmetic formulas. For this reason, combinatorics was closely integrated with the theory of numbers for two centuries.

In the twentieth century, new applications appeared: in chemistry (Pólya proved his famous theorem in order to count polymers), in physics (the duality problem), in economics and operational research (the traveling salesman problem), in statistics (the design of experiments), in information theory (the problem of the capacity of a set of signals), etc.

When the configurations being studied become too complex, it proved futile to attempt to connect them with known algebraic structures and to count them with "elegant formulas"; rather than equalities, it was necessary to resort to inequalities, asymptotic values, congruences mod p, etc.

Instead of trying to find the precise number of configurations with a given property, one looked for information about this number without necessarily giving exact, or even recurrence, formulas for finding it. A particularly curious case in point being the Ramsey numbers, so close to the binomial coefficients, yet so elusive that it is not even known how to calculate them when the values of the parameters are greater than 7.

FIFTH ASPECT: ENUMERATION OF CONFIGURATIONS

If one is interested in finding the number of configurations satisfying certain given conditions then this is a *counting* problem; if one wants to actually list these configurations, then it is a problem in *enumeration* (these words usually have the same meaning, but we shall retain this distinction).

Actually enumerating configurations is usually an unrewarding task as well as often being beyond the capacity of not only any one human being, but even the fastest computers; nevertheless, for certain particularly difficult problems, it is still the only method (reasoning by exhaustion) of proof. Some important results in topology, for example, have been discovered by starting with lists of convex polyhedra.

The problem of enumeration is inseparable from the problem of classification; if the set of configurations under consideration is divided into classes, then it is sometimes sufficient to prove the required result for just one member of each class.

In very simple cases (for example, combinations of n objects taken p at a time) a counting formula is obtained by determining, in the first place, a method for enumeration; but the converse procedure can be equally productive: Arithmetic operations (on the numbers) induce algebraic operations (on the set of configurations), and a method of counting becomes a procedure for generating configurations. This is what happens [8] when one enumerates the paths of a given length in a graph; probably much of classical combinatorics could be rewritten from this viewpoint.

SIXTH ASPECT: OPTIMIZATION

The following problem is typical of the problems encountered in operational research.

Suppose each configuration x, with certain specified properties, is assigned a numerical value $f(x)$ (f is called the economic function). The problem is to choose a configuration x_0 which minimizes $f(x)$, or makes it ε-minimal, i.e.,

$$f(x) \geqslant f(x_0) - \varepsilon$$

for all configurations x with the specified properties.

The celebrated "traveling salesman problem" is an example of this type: find the shortest itinerary for a salesman if he wishes to visit once, and only once, all 50 state capitals of the United States and return to his point of departure. Up to the present there have been no really satisfactory solutions to this apparently innocuous problem. The suggested algorithms for finding the optimal itinerary involve far too many steps (for a "good" algorithm, the order of magnitude of the number of steps must be a polynomial in n; beyond that, the calculations very quickly become unmanageable, even for a computer).

In very delicate cases, approximate procedures, such as "SEP" [9], or "Branch and Bound," must be used (the Little *et al.* [10] approach to the traveling salesman problem; the Land and Doig [11] method for solving linear programming problems, etc.).

These procedures consist, roughly speaking, of considering progressively smaller and smaller subsets of acceptable solutions until it can be shown that one of them contains an ε-optimal solution (where ε is sufficiently small). Besides these types of optimization (or "maximization") problems, there are others that involve choosing a set of acceptable configurations according to certain well-defined criteria. Some are, for example, multiple-choice problems (for instance, Electre's [12] method); the problem of finding a "solution" to a noncooperative game (in the von Neumann–Morgenstein sense): the problem of finding the "kernel" of a graph. In general, in this sort of problem, it is not just a question of choosing one optimal configuration but a set of configurations, satisfying certain criteria, from which a further choice can be made.

* * *

Considering all these aspects of combinatorics, one is struck by the recent proliferation of problems that are expressible in terms of configurations, and by the diversity of methods for solving them.

A great number of combinatorial problems can be studied in the framework of a coherent and well-established mathematical theory (theory of graphs, Galois theory, and Boolean algebras). However, there exist many other problems not in this category and the methods used for solving them very urgently need to be unified, formalized, and generalized.

The present work is concerned only with the problems of counting, that is, the oldest preoccupation of combinatorics. This is the text of a series of lectures given in the *Faculté des Sciences de Paris* in 1967–1968 in the framework of the *Maîtrise de Mathématiques et Applications Fondamentales*. Throughout these lectures, modern set-theoretic terminology is used rather than the older and somewhat ambiguous terminology (for example, "combinations with repetitions," etc.). In addition, instead of citing the classical applications in the theory of special functions or in number theory, we have systematically preferred those applications that are either of more general interest, or that arise from new fields of study, such as information theory.

Finally, we have systematically avoided the use of symbolic calculus which, although it is sometimes the most convenient tool for manipulating very complicated formulas, is generally used without sufficient justification. One assumes that a property that can be interpreted as an equality $|A| = |B|$ is better expressed by constructing a bijection between the sets A and B rather than calculating the coefficients of a polynomial, the variables of which have no particular significance.*

REFERENCES

1. R. JAULIN, "La Géomancie, Cahiers de l'Homme." Mouton, Paris-La Haye, 1966.
2. I. THOMAS, "Selections Illustrating the History of Greek Mathematics," 2 vols. Harvard Univ. Press, Cambridge, Mass., 1939.

* The method of generating functions, having played havoc for a century, has fallen into disuse for this reason; in any case, it should not be used outside the framework of the ring of formal power series (see, for example, P. Dubreil and M. L. Dubreil-Jacotin [13]).

3. "Sacred Books of the East" (F. M. Müller, ed.), Vol. XVI ("The Yi-King"). Oxford Univ. (Clarendon) Press, London, 1882.

4. F. HARARY, *Publ. Math. Inst. Hungarian Acad. Sci.* **5**, 92–93 (1960).

5. J. H. REDFIELD, The theory of group-reduced distributions, *Amer. J. Math.* **49**, 433–455 (1927).

6. J. RIORDAN, "An Introduction to Combinatorial Analysis." Wiley, New York, 1958.

7. NASIR-AD-DIN AT-TUSI, "Handbook of Arithmetic Using Board and Dust" (Russian translation by S. A. Ahmedov and B. A. Rosenfeld); *Istor. Mat. Issled.* **15**, 431–444 (1963); *Math. Rev.*, **31**, 5776 (1966).

8. A. KAUFMANN and Y. MALGRANGE, Recherche des Chemins et Circuits hamiltoniens d'un graphe, *Rev. Française Informat. Recherche Operationnelle* **26** (1963); see also A. Kaufmann, "Initiation à la Combinatorique." Dunod, Paris, 1968.

9. See P. BERTIER and B. ROY, Une procédure de resolution pour une classe de problèmes pouvant avoir un caractère combinatoire, *ICC Bull.* **4**, 19–28 (1965).

10. J. D. C. LITTLE, K. G. MURTY, D. W. SWEENEY, and C. KAREL, An algorithm for the traveling salesman problem, *Operations Res.* **11** (1963).

11. A. G. DOIG and A. H. LAND, An automatic method for solving discrete programming problems, *Econometrica* **28** (1960).

12. P. BUFFET, J. P. GREMY, M. MARC, and B. SUSSMAN, Peut-on choisir en tenant compte de critères multiples, *Metra* **6**, (1967).

13. P. DUBREIL and M. L. DUBREIL-JACOTIN, "Leçons d'Algèbre Moderne," 2nd ed., /61, pp. 122–133. Dunod, Paris, 1964.

14. R. C. BOSE, S. S. SHRIKHANDE, and E. T. PARKER, Further results on the construction of mutually orthogonal Latin squares and the falsity of Euler's conjecture, *Canad. J. Math.* **12**, 189–203 (1960).

CHAPTER 1

THE ELEMENTARY COUNTING FUNCTIONS

1. MAPPINGS OF FINITE SETS

A *finite set* A is defined to be a collection of distinct objects a_1, $a_2, \ldots, a_m$. Standard notation is as follows:

$a \in A$	a is an element of the set A,
$\lvert A \rvert$	cardinality of A (the number of elements of the finite set A),
$A \subset B$	A is a subset of B (each element of A is an element of B),
$A \cup B$	the union of the sets A and B (the set of elements belonging to either A or B),
$A \cap B$	the intersection of the sets A and B (the set of elements common to A and B),
$A = X - A$	the complement of A relative to X,
$\varnothing$	the empty set (no elements),
$\mathscr{P}(A)$	the family of subsets of A (including $\varnothing$ and A),
$A \times B$	the Cartesian product of the sets A and B (the set of ordered pairs (a, b), $a \in A$, $b \in B$),

$$A^n = A \times A \times \cdots \times A \quad \text{the set of } n\text{-tuples } (a_1, a_2, \ldots, a_n),\ a_i \in A,\ i = 1, 2, \ldots, n.$$

Let A and X be sets; denote the elements of A by $a_1, a_2, \ldots, a_m$ and those of X by $1, 2, \ldots, n$. A *mapping* of X into A is a rule, denoted by

$$\varphi = \begin{pmatrix} 1 & 2 & \cdots & n \\ a_{i_1} & a_{i_2} & \cdots & a_{i_n} \end{pmatrix},$$

which associates the object 1 of the set X with the object $a_{i_1} = \varphi(1)$ of the set A, the object 2 with the object $a_{i_2} = \varphi(2)$, etc.

FIRST INTERPRETATION: ARRANGING n OBJECTS. We may sometimes regard
 X as a set of objects to be sorted into boxes;
 A as the set of boxes.
Then, each *way of sorting* the objects into the boxes is a mapping of X into A.

EXAMPLE. The mapping

$$\varphi = \begin{pmatrix} 1 & 2 & 3 & 4 & 5 \\ c & a & b & b & a \end{pmatrix}$$

represents the arrangement
 objects 2 and 5 in box a,
 objects 3 and 4 in box b,
 object 1 in box c.

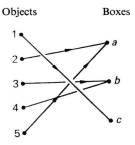

SECOND INTERPRETATION: n-TUPLES OF LETTERS OR "WORDS" OF LENGTH n.
 X is a set of numbered places $1, 2, \ldots, n$;
 A is a set of letters to be arranged in these places.
 Then a mapping of X into A is an n-tuple of letters (a letter may occur more than once).

EXAMPLE. The mapping

$$\varphi = \begin{pmatrix} 1 & 2 & 3 & 4 & 5 \\ c & a & b & b & a \end{pmatrix}$$

represents the word *cabba* and is sometimes written as

$$\varphi = cabba.$$

Thus, there is a *duality* between the arrangements of n objects and the words of n letters which will often be utilized below.

A mapping φ is *surjective* (or a *surjection*) if, for every element $a \in A$, there exists at least one $i \in X$ such that $a = \varphi(i)$ (then the arrangement has at least one object in every box; the word contains all the letters).

A mapping φ is *injective* (or an *injection*) if

$$i \neq j \quad \text{implies} \quad \varphi(i) \neq \varphi(j)$$

(then the arrangement has at most one object in each box; the word consists of distinct letters).

A mapping is *bijective* (or a *bijection*) if it is both injective and surjective (then the arrangement consists of exactly one object in each box; the word contains each of the letters once and only once; it is a "permutation" of the n letters).

EXAMPLE. $X = \{1, 2, \ldots, 7\}$, $A = \{a, b, c, d, e\}$. The mapping

$$\varphi = \begin{pmatrix} 1 & 2 & 3 & 4 & 5 & 6 & 7 \\ a & c & e & a & e & b & d \end{pmatrix} = aceaebd$$

is surjective, implying $|A| \leqslant |X|$.

If $X = \{1, 2, 3\}$, $A = \{a, b, c, d, e\}$, the mapping

$$\varphi = \begin{pmatrix} 1 & 2 & 3 \\ d & b & e \end{pmatrix} = dbe$$

is injective, implying that $|X| \leqslant |A|$.

Note that the mappings of X into A, *which have as large a range of values in A as possible,* are necessarily

injective if $|X| < |A|$,

bijective if $|X| = |A|$,

surjective if $|X| > |A|$.

A simple way of showing that two sets have the same number of elements is to define a bijection between them.

2. The Cardinality of the Cartesian Product $A \times X$

Consider the set A of m elements $a_1, a_2, \ldots, a_m$ and the set X of n elements $1, 2, \ldots, n$, then $A \times X$ has exactly $m \times n$ elements. Symbolically:

PROPOSITION 1. $|A \times X| = |A| \times |X|$.

APPLICATION (Erdös, Szekeres [4]). *A sequence of $mn + 1$ distinct integers $u_1, u_2, \ldots, u_{mn+1}$ contains either a decreasing subsequence of length greater than m or an increasing subsequence of length greater than n.*

PROOF: Let l_i^- be the length of the longest decreasing subsequence with first term u_i and l_i^+ be the length of the longest increasing subsequence with first term u_i.

Assume that the proposition is false. Then $u_i \to (l_i^-, l_i^+)$ defines a mapping of $\{u_1, u_2, \ldots, u_{mn+1}\}$ into the Cartesian product $\{1, 2, \ldots, m\} \times \{1, 2, \ldots, n\}$; this mapping is injective since, if $i < j$,

$$u_i > u_j \Rightarrow l_i^- > l_j^- \Rightarrow (l_i^-, l_i^+) \neq (l_j^-, l_j^+),$$

$$u_i < u_j \Rightarrow l_i^+ > l_j^+ \Rightarrow (l_i^-, l_i^+) \neq (l_j^-, l_j^+).$$

Hence, by Proposition 1,

$$mn + 1 \leqslant mn,$$

which is impossible.

Note that this property has frequently been rediscovered; Frasnay [5] generalized it in the following way.

In a sequence $(u_1, u_2, \ldots)$ of $n_1 n_2 \cdots n_p + 1$ distinct elements we are given p total order relations $\overset{1}{<}, \overset{2}{<}, \ldots, \overset{p}{<}$ such that the sets $R_k = \{(u_i, u_j)/u_i \overset{k}{<} u_j\}$ cover the set $\{(u_i, u_j)/u_i \neq u_j\}$; then there exists a relation $\overset{k}{<}$ and a subsequence $(u_{i_1}, u_{i_2}, \ldots)$ of length $n_k + 1$, such that

$$u_{i_1} \overset{k}{<} u_{i_2} \overset{k}{<} \cdots \qquad (1 \leqslant k \leqslant p).$$

This can be proved in exactly the same way as above, by using the relation

$$|A_1 \times A_2 \times \cdots \times A_p| = |A_1| \times |A_2| \times \cdots \times |A_p|.$$

3. NUMBER OF SUBSETS OF A FINITE SET A

Let us try to count the number $|\mathscr{P}(A)|$ of subsets of the set $A = \{a_1, a_2, \ldots, a_m\}$.

EXAMPLE. $A = \{a, b, c, d\}$.
The subsets of A are listed in the following diagram:

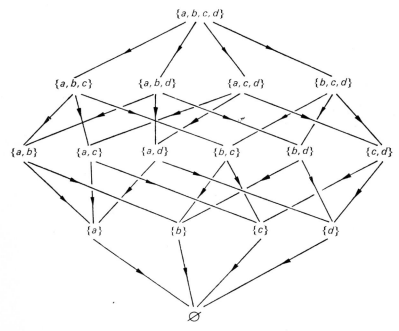

In this figure a line is directed from the set S to the set T if

(1) $S \supset T$;
(2) $S \supset R \supset T$ *implies* $R = S$ or $R = T$.

Since all the arrows are directed downwards, it is usual to omit them. Obviously $\supset$ is an order relation, i.e.,

(1) $S \supset S$;
(2) $S \supset T, T \supset S$ *implies* $S = T$;
(3) $S \supset T, T \supset U$ *implies* $S \supset U$.

   Moreover, for any two elements S and T there exist a least upper bound $S \cup T$ and a greatest lower bound $S \cap T$. In other words, $\mathscr{P}(A)$ is a *lattice*.
   Note that

$$|\mathscr{P}(A)| = 16 = 2^4.$$

   REMARK: The subsets of a set of m elements can always be represented by the vertices of a hypercube in m-dimensional space.

For example, with $A = \{a, b, c\}$,

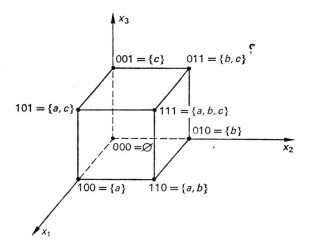

In this representation the edges of the hypercube are the same as the edges (unorientated) in the preceding representation. In this case

$$|\mathscr{P}(A)| = 8 = 2^3.$$

PROPOSITION. *The number of subsets of a set A with m elements is*

$$|\mathscr{P}(A)| = 2^m.$$

PROOF: The proof is obvious.

4. NUMBERS m^n, OR MAPPINGS OF X INTO A

Let us consider a set A of m elements and the Cartesian product

$$A \times A \times \cdots \times A = A^n;$$

i.e., the set of *n-tuples* $(a_1, a_2, \ldots, a_n)$ with $a_1, a_2, \ldots, a_n \in A$.

EXAMPLE. The 4-tuples which can be formed with the letters a and b are

$$\begin{array}{cccc}
aaaa & abaa & baaa & bbaa \\
aaab & abab & baab & bbab \\
aaba & abba & baba & bbba \\
aabb & abbb & babb & bbbb.
\end{array}$$

There are $16 = 2^4$ in all.

PROPOSITION. *The number of n-tuples on m letters is equal to m^n.*

PROOF: An *n*-tuple contains n letters: there are m possibilities for the first letter; m possibilities for the second letter; etc. Hence, the number of such *n*-tuples is $m \times m \times \cdots \times m = m^n$.

EQUIVALENT PROPOSITION. *The number of mappings of a set X of n elements into a set A of m elements is equal to m^n.*

PROOF: This follows immediately from the correspondence

$$\begin{pmatrix} 1 & 2 & \cdots & n \\ a_{i_1} & a_{i_2} & \cdots & a_{i_n} \end{pmatrix} \to a_{i_1} a_{i_2} \cdots a_{i_n}.$$

5. NUMBERS $[m]_n$, OR INJECTIONS OF X INTO A

If x is a real number, let

$$[x]_n = \underbrace{x(x-1)(x-2) \cdots (x-n+1)}_{n}.$$

Let m and n be integers such that $m \geqslant n$; now we shall count the number of n-tuples of letters from the set A ($|A| = m$) with the restriction that no letter be repeated.

EXAMPLE. $A = \{a, b, c, d\}$.

With $m = 4$, $n = 2$ there are twelve 2-tuples:

$$\left.\begin{array}{cccc} ab & ba & ca & da \\ ac & bc & cb & db \\ ad & bd & cd & dc \end{array}\right\} (T_2).$$

PROPOSITION. *The number of n-tuples without repetitions on m letters is equal to $[m]_n$.*

PROOF: Form the table T_{n-1} of all the $(n-1)$-tuples without repetitions; then form the table T_n by successively adding to the end of every word of T_{n-1} each of the $(m - n + 1)$ letters that does not appear in that word.

In this way, we obtain the table of all the n-tuples without repetitions, since

(1)   these are n-tuples without repetitions;

(2)   there are no omissions; the n-tuple $(\alpha_1, \alpha_2, \ldots, \alpha_n)$ appears in T_n since necessarily $(\alpha_1, \alpha_2, \ldots, \alpha_{n-1})$ appears in T_{n-1};

(3)   there are no repetitions, since either two n-tuples of T_n differ in the first $(n-1)$ letters if they come from different $(n-1)$-tuples, or in the nth letter if they come from the same $(n-1)$-tuple.

Therefore,

$$\begin{aligned} |T_n| &= (m - n + 1)|T_{n-1}|, \\ &= (m - n + 1)(m - n + 2) \cdots (m - n)|T_1|, \\ &= (m - n + 1)(m - n + 2) \cdots (m - 1)m = [m]_n. \end{aligned}$$

PROPOSITION. *The number of injections of a set X of n elements into a set A of m elements is equal to $[m]_n$.*

PROOF: An n-tuple without repetitions, such as $adcb$, corresponds to an injective mapping

$$\varphi = \begin{pmatrix} 1 & 2 & 3 & 4 \\ a & d & c & b \end{pmatrix}.$$

m!, OR "FACTORIAL *m*." It is usual to write

$$m! = \begin{cases} 1 \times 2 \times \cdots \times m & \text{if the integer } m \\ & \text{is greater than 0,} \\ 1 & \text{if } m = 0. \end{cases}$$

m! *is, therefore, the number of permutations of m objects* $a_1, a_2, \ldots, a_m$, that is, the number of ways of arranging *m* distinct objects in *m* given places.

Thus,

$$[m]_n = \frac{m(m-1)(m-2)\cdots(m-n+1)(m-n)\cdots 1}{(m-n)\cdots 1} = \frac{m!}{(m-n)!}.$$

STIRLING NUMBERS OF THE FIRST KIND. $[x]_n$, being a polynomial of degree *n*, can be expressed as

$$[x]_n = s_n^{\ 0} + s_n^{\ 1} x + s_n^{\ 2} x^2 + \cdots + s_n^{\ n} x^n$$

By definition, the coefficients $s_n^{\ k}$ are the *Stirling numbers of the first kind.*

RECURRENCE RELATIONS. The Stirling numbers of the first kind can be calculated from the following formulas:

$$s_{n+1}^k = s_n^{k-1} - n s_n^{\ k},$$
$$s_n^{\ 0} = 0,$$
$$s_n^{\ n} = 1.$$

PROOF: By definition,

$$[x]_{n+1} = [x]_n(x - n)$$

and therefore, again by definition,

$$\cdots + s_{n+1}^k x^k + \cdots = (\cdots + s_n^{k-1} x^{k-1} + s_n^{\ k} x^k + \cdots)(x - n).$$

Equating the coefficients of x^k on the left- and right-hand sides of this equation gives the first formula above. The other two formulas are obvious.

The following values can be obtained recursively from these formulas:

s_n^k	$k = 0$	1	2	3	4 $\cdots$
$n = 1$	0	1	0	0	0 $\cdots$
2	0	-1	1	0	0 $\cdots$
3	0	2	-3	1	0 $\cdots$
4	0	-6	11	-6	1 $\cdots$

6. NUMBERS $[m]^n$

If x is a real number, let

$$[x]^n = x(x + 1)(x + 2) \cdots (x + n - 1).$$

Suppose that X is a set of n objects $1, 2, \ldots, n$ to be arranged in boxes, each box may contain any number of objects, these being placed in the box in a given order ("ordered boxes").

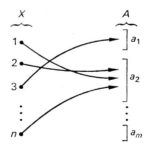

We shall now find the number of arrangements of n objects in m ordered boxes.

EXAMPLE. Consider arrangements in two ordered boxes; we shall denote the case, for example, where the objects i, j, k are in the first box (in that order), and the object l is in the second box by ijk/l; when $X = \{1, 2\}$, the possible arrangements are

$$\emptyset/12 \quad 1/2 \quad 12/\emptyset$$
$$\emptyset/21 \quad 2/1 \quad 21/\emptyset$$

PROPOSITION. *The number of ways of arranging n objects in m ordered boxes is equal to $[m]^n$.*

PROOF: First, form the table T_{n-1} of all the arrangements of objects $1, 2, \ldots, n-1$ in the m ordered boxes. Each arrangement

$$i_1 \, i_2 \, \cdots \, /i_k \, i_{k+1} \, \cdots \, / \, \cdots \, / \, \cdots \, i_{n-1}$$

can be expressed as a sequence of $(n-1) + (m-1)$ symbols (either letters i_k or sloping lines $/$); the letter n can be added to this sequence in $(n-1) + (m-1) + 1$ different ways.

Thus,

$$|T_n| = (m + n - 1)|T_{n-1}|$$
$$= (m + n - 1)(m + n - 2) \cdots (m + 1)|T_1| = [m]^n.$$

7. NUMBERS $[m]^n/n!$, OR INCREASING MAPPINGS OF X INTO A

Let A be the set of letters $a_1, a_2, \ldots, a_m$ ordered so that

$$a_1 < a_2 < a_3 < \cdots < a_m.$$

A word $x_1 \, x_2 \, \cdots \, x_n$ of length n is *increasing* if

$$x_1 \leqslant x_2 \leqslant x_3 \leqslant \cdots \leqslant x_n.$$

EXAMPLE. Let $A = \{a, b, c, d\}$, where $a < b < c < d$. The increasing words of length 3 are

$$
\begin{array}{llll}
aaa & abb & acc & add \\
aab & abc & acd \\
aac & abd \\
aad \\
\\
bbb & bcc & bdd \\
bbc & bcd \\
bbd \\
\\
ccc & cdd \\
ccd \\
\\
ddd
\end{array}
$$

20 in all. (In old-fashioned terminology, the above elements are called "combinations with repetitions of m objects taken n at a time.")

PROPOSITION. *The number of increasing words of length n on m letters is equal to $[m]^n/n!$.*

PROOF: Consider an arrangement of n objects $1, 2, \ldots, n$ in m ordered boxes $a_1, a_2, \ldots, a_m$, as in the preceding paragraph, and let it correspond to an increasing word in the following way:

$$| \quad 3 \quad | \quad | \quad 251 \quad | \quad | \quad \quad | \quad \quad | \quad | \quad 647 \quad | \qquad \rightarrow a_1 a_2 a_2 a_2 a_4 a_4 a_4 .$$
$$\underbrace{\qquad}_{a_1} \quad \underbrace{\qquad}_{a_2} \quad \underbrace{\qquad}_{a_3} \quad \underbrace{\qquad}_{a_4}$$

The letter a_1 is written as often as the number of objects in a_1, followed by the letter a_2 written as often as the number of objects in a_2, etc.

To each arrangement of n objects there corresponds one and only one increasing word; an increasing word, on the other hand, corresponds to exactly $n!$ distinct arrangements. Therefore, by the previous proposition, the number of increasing words is $[m]^n/n!$.

EQUIVALENT PROPOSITION. *The number of increasing mappings of X into A is equal to $[m]^n/n!$.*

APPLICATION (De Moivre). Let us determine the number of ways of obtaining the integer m as a summation of n nonnegative integers, i.e., $m = u_1 + u_2 + \cdots + u_n$. A solution $u_1', u_2', \ldots, u_n'$ is the same as the solution $u_1, u_2, \ldots, u_n$ if, and only if, $u_1' = u_1, u_2' = u_2$, etc.

Suppose $s_k = u_1 + u_2 + \cdots + u_k$. Each summation is uniquely defined by a word $s_1 s_2 \cdots s_{n-1}$, where

$$0 \leqslant s_1 \leqslant s_2 \leqslant \cdots \leqslant s_{n-1} \leqslant m.$$

Therefore, the number of solutions is

$$\frac{[m+1]^{n-1}}{(n-1)!} = \frac{(m+n-1)!}{(n-1)!m!} .$$

8. BINOMIAL NUMBERS

Let us determine the number of subsets of n elements of a set A with cardinality m.

EXAMPLE. $A = \{a, b, c, d, e\}$, $n = 3$. The subsets are

$$
\begin{array}{lll}
abc & acd & ade \\
abd & ace & \\
abe & & \\[4pt]
bcd & bde & \\
bce & & \\
cde & &
\end{array}
$$

There are 10 subsets of A with 3 elements. This table may also be considered as the list of *strictly increasing* words of length 3.

PROPOSITION. *The number of subsets of A with n elements is*

$$
\frac{[m]_n}{n!} = \frac{m(m-1)(m-2)\cdots(m-n+1)}{1\cdot 2\cdots n} = \frac{m!}{n!(m-n)!}.
$$

PROOF: Form the table T of strictly increasing n-tuples from the set A of m letters; permute the letters in each of these n-tuples in every possible way and let T' be the set of resulting n-tuples.

Thus, T' is the set of n-tuples without repetitions with elements from A. There are no omissions; every n-tuple of distinct letters appears in T'. There are no repetitions, because two n-tuples of T' are derived, either from the same n-tuple of T and so differ in the order of the letters, or from two different n-tuples of T in which case they are formed by different letters.

By Section 5, T' has $[m]_n$ elements and so T has $[m]_n/n!$ elements. **Q.E.D.**

We write

$$
\binom{m}{n} = \begin{cases} \dfrac{[m]_n}{n!} & \text{if} \quad n \neq 0, \quad m \geqslant n, \\[8pt] 1 & \text{if} \quad n = 0, \quad m \geqslant n, \\[8pt] 0 & \text{if} \quad m < n. \end{cases}
$$

$\binom{m}{n}$ is called the *binomial number of m with respect to n*. In old-fashioned terminology it is called "the number of combinations of m objects taken n at a time" and is sometimes denoted by $C_m{}^n$. (Note the upper index in one symbol is the lower index in the other.)

FORMULA 1. *For $m \geqslant n > 0$,*

$$
\binom{m}{n} = \binom{m-1}{n} + \binom{m-1}{n-1}
$$
$$
\binom{m}{0} = \binom{m}{m} = 1
$$

PROOF: Form the table T of strictly increasing n-tuples from $A = \{a_1, a_2, \ldots, a_m\}$; T can be partitioned into two tables, T' consisting of n-tuples containing a_1, and T'' consisting of n-tuples not containing a_1. T' has exactly the same number of elements as the table of $(n-1)$-tuples from $\{a_2, a_3, \ldots, a_m\}$ and T'' that of n-tuples from $\{a_2, a_3, \ldots, a_m\}$; hence, we get the first equality. The other two equalities are obvious.

REMARK: Using Formula 1, all the numbers $\binom{m}{n}$ can be calculated recursively, e.g.,

$\binom{m}{n}$	$n=0$	$n=1$	$n=2$	$n=3$	$n=4$	$n=5$	$\cdots$
$m=1$	1	1	0	0	0	0	$\cdots$
$m=2$	1	2	1	0	0	0	$\cdots$
$m=3$	1	3	3	1	0	0	$\cdots$
$m=4$	1	4	6	4	1	0	$\cdots$
$m=5$	1	5	10	10	5	1	$\cdots$
$\vdots$							

This table is called *Pascal's triangle*. Notice the following.

(1) Considering the nonzero entries of each row only, the coefficients equidistant from the ends are equal, that is

$$\boxed{\binom{m}{n} = \binom{m}{m-n}}$$

Indeed, both coefficients are equal to

$$\frac{m!}{n!(m-n)!}.$$

(2) For fixed m the numbers $\binom{m}{n}$ increase with n until

$$n \leqslant (m+1)/2.$$

Indeed,

$$\binom{m}{n-1} \leqslant \binom{m}{n}$$

implies

$$\frac{m!}{(n-1)!(m-n+1)!} \leqslant \frac{m!}{n!(m-n)!},$$

or

$$\frac{1}{m-n+1} \leqslant \frac{1}{n},$$

or

$$n \leqslant \frac{m+1}{2}.$$

FORMULA 2 (Generalization of Formula 1). *For* m, n, $p \geqslant 1$, $m+p \geqslant n$,

$$\boxed{\binom{m+p}{n} = \sum_{k=0}^{m} \binom{m}{k}\binom{p}{n-k}}$$

REMARK: For $p = 1, n \geqslant 1$, we have

$$\binom{m+1}{n} = \binom{m}{n-1} + \binom{m}{n}.$$

This is Formula 1.

For $p = 2, n \geqslant 2$, the formula gives

$$\binom{m+2}{n} = \binom{m}{n-2} + 2\binom{m}{n-1} + \binom{m}{n}.$$

More generally, if $n \geqslant p$, one obtains

$$\binom{m+p}{n} = \binom{m}{n-p} + \binom{m}{n-p+1}\binom{p}{p-1}$$

$$+ \binom{m}{n-p+2}\binom{p}{p-2} + \cdots + \binom{m}{n}\binom{p}{0}. \quad (1)$$

If $n \leqslant p$, one obtains

$$\binom{m+p}{n} = \binom{m}{0}\binom{p}{n} + \binom{m}{1}\binom{p}{n-1} + \cdots + \binom{m}{n}\binom{p}{0}. \quad (2)$$

PROOF OF FORMULA 2: Form the table T of subsets of $\{a_1, a_2, \ldots, a_m, b_1, b_2, \ldots, b_p\}$ with n elements; assuming $n \geqslant p$: $\binom{m}{n-p}$ of these subsets contain all the b_i's; $\binom{p}{p-1}\binom{m}{n-p+1}$ of these subsets contain $p-1$ of the b_i's, etc. Summing, we obtain Eq. (1).

Similarly, if $n \leqslant p$, we obtain Eq. (2).

THE BINOMIAL FORMULA IN A COMMUTATIVE RING. Consider a set A together with a binary operation $+$, called *addition*, satisfying the following axioms:

(1) $a + (b + c) = (a + b) + c$ (associativity);

(2) $a + b = b + a$ (commutativity);

(3) the existence of a neutral element 0 satisfying $0 + a = 0$ for all $a \in A$;

(4) the existence of an inverse element $-a$, for each $a \in A$, satisfying $a + (-a) = 0$.

Then A is said to be an *Abelian group*; suppose now that there exists another binary operation, $\cdot$, called multiplication, defined on A and satisfying the following axioms:

(5) $a \cdot (b \cdot c) = (a \cdot b) \cdot c$ (associativity);
(6) $a \cdot b = b \cdot a$ (commutativity);
(7) $a \cdot (b + c) = a \cdot b + a \cdot c$ (distributivity with respect to addition).

Then A is called a *commutative ring*.

EXAMPLES. The set Z of integers; the set R of real numbers; the set Γ of rational numbers; the set Ω of complex numbers; the set of polynomials with coefficients in Z, R, Γ, or Ω; the set of bounded real functions defined on the interval $[0, 1]$, etc.

Let us consider two n-tuples $(a^1, a^2, \ldots, a^n)$ and $(b^1, b^2, \ldots, b^n)$ of elements of A and form the product

$$(a^1 + b^1)(a^2 + b^2) \cdots (a^n + b^n) = \prod_{i=1}^{n} (a^i + b^i).$$

If $K \subset N = \{1, 2, \ldots, n\}$, and if $|K| = k \neq 0$, write

$$a^K = \prod_{i \in K} a^i, \qquad b^K = \prod_{i \in K} b^i.$$

If $k = 0$, let $a^{\varnothing} b^N = b^N$, $a^N b^{\varnothing} = a^N$; then

$$\prod_{i \in N} (a^i + b^i) = (a^1 + b^1)(a^2 + b^2) \cdots (a^n + b^n) = \sum_{K \subset N} a^K b^{N-K}.$$

Letting $a^1 = a^2 = \cdots = a^n = a$, $b^1 = b^2 = \cdots = b^n = b$, we obtain

$$\boxed{(a + b)^n = \sum_{k=0}^{n} \binom{n}{k} a^k b^{n-k}}$$

APPLICATION. Consider, in the vector space Φ of real functions, a linear transformation P, i.e., a rule P, which associates with every function $\varphi \in \Phi$ a function $P\varphi \in \Phi$ such that

$$P[\varphi(x) + \psi(x)] = P\varphi(x) + P\psi(x),$$

$$P[\lambda\varphi(x)] = \lambda P\varphi(x), \qquad \lambda = \text{constant.}$$

If P and Q are two linear transformations, then $P + Q$ and PQ are defined by

$$(P + Q)\varphi(x) = P\varphi(x) + Q\varphi(x),$$

$$(PQ)\varphi(x) = P[Q\varphi(x)].$$

Now consider two linear transformations P and Q that commute (i.e., $PQ = QP$); the set of linear transformations generated by P and Q (by repeated addition and multiplication) forms a commutative ring; thus,

$$(P + Q)^n\varphi(x) = \sum_{k=0}^{n} \binom{n}{k} P^k Q^{n-k}\varphi(x).$$

In particular, consider the linear transformations

$$E\varphi(x) = \varphi(x + 1),$$
$$I\varphi(x) = \varphi(x),$$
$$\Delta\varphi(x) = \varphi(x + 1) - \varphi(x) = (E - I)\varphi(x).$$

Then,

$$\Delta^n\varphi(x) = (E - I)^n\varphi(x) = \sum_{k=0}^{n} (-1)^{n-k} \binom{n}{k} E^k\varphi(x).$$

Finally we obtain the often used formula

$$\Delta^n\varphi(x) = \sum_{k=0}^{n} (-1)^{n-k} \binom{n}{k} \varphi(x + k)$$

FIBONACCI NUMBERS

(1)   We will now try to determine the number $f(n, k)$ of subsets of $X = \{1, 2, \ldots, n\}$ having k elements, no two of which are consecutive integers.

Associate with each such subset S a word $\alpha_1, \alpha_2, \ldots, \alpha_n$, where

$$\alpha_i = \begin{cases} 0 & \text{if } i \notin S, \\ 1 & \text{if } i \in S. \end{cases}$$

In this word, consisting of 0's and 1's, no two 1's will be adjacent. This mapping between such subsets S and such words $\alpha_1, \alpha_2, \ldots, \alpha_n$ is bijective; hence, instead of counting subsets, we may count the number of such words.

Consider $n - k$ digits 0, numbered from 1 to $n - k$, and interpose k digits 1 in such a way that no two of them are adjacent; each 1 can be characterized by the number given to the 0 immediately preceding it: hence, k distinct integers must be chosen from the set $\{0, 1, 2, \ldots, n - k\}$, giving a total number of possibilities equal to

$$f(n, k) = \binom{n - k + 1}{k}.$$

The number of subsets of X not containing two consecutive integers is therefore

$$F_n = \sum_k \binom{n - k + 1}{k},$$

where F_n is a *Fibonacci number*.

(2) Let us try to determine the number $f^*(n, k)$ of subsets of X, with k elements, which do not contain either two consecutive integers or both 1 and n simultaneously.

The subsets which contain n cannot contain either $n - 1$ or 1, and hence there are $f(n - 3, k - 1)$ of them; those not containing n number $f(n - 1, k)$. It follows that

$$f^*(n, k) = f(n - 3, k - 1) + f(n - 1, k)$$

$$= \binom{n - k - 1}{k - 1} + \binom{n - k}{k}$$

$$= \binom{n - k}{k}\left[\frac{k}{n - k} + 1\right]$$

$$= \frac{n}{n-k} \binom{n-k}{k}.$$

The number of subsets of X that do not contain two consecutive integers modulo $(n-1)$ is, therefore,

$$F_n^* = \sum_k \frac{n}{n-k} \binom{n-k}{k},$$

where F_n^* is sometimes called a *corrected Fibonacci number*.

9. Multinomial Numbers $\binom{n}{n_1, n_2, \ldots, n_p}$

Proposition. *Let X be a set of objects and let $n_1, n_2, \ldots, n_p$ be nonnegative integers satisfying $n_1 + n_2 + \cdots + n_p = n$; the number of arrangements of the objects into boxes $X_1, X_2, \ldots, X_p$, each containing $n_1, n_2, \ldots, n_p$ objects, respectively, is*

$$\binom{n}{n_1, n_2, \ldots, n_p} = \begin{cases} \dfrac{n!}{n_1! n_2! \cdots ! n_p!} & \text{if } n_1 + n_2 + \cdots + n_p = n, \\ 0 & \text{otherwise.} \end{cases}$$

Proof: Let $n_1 + n_2 + \cdots + n_p = n$. The box X_1 can be filled in $\binom{n}{n_1}$ different ways; now, assuming X_1 is filled, X_2 can be filled in $\binom{n-n_1}{n_2}$ different ways, etc.

Hence, the required number is

$$\binom{n}{n_1}\binom{n-n_1}{n_2}\binom{n-n_1-n_2}{n_3} \cdots \binom{n_p}{n_p}$$

$$= \frac{n!}{n_1!(n-n_1)!} \frac{(n-n_1)!}{n_2!(n-n_1-n_2)!} \frac{(n-n_1-n_2)!}{n_3!(n-n_1-n_2-n_3)!} \cdots \frac{n_p!}{n_p!}$$

$$= \frac{n!}{n_1! n_2! n_3! \cdots n_p!}.$$

THE MULTINOMIAL FORMULA IN A COMMUTATIVE RING. If A is a commutative ring (see Section 8) and if $a_1, a_2, \ldots, a_p \in A$, then

$$(a_1 + a_2 + \cdots + a_p)^n = \sum_{\substack{n_1, n_2, \ldots, n_p \geqslant 0 \\ n_1 + n_2 + \cdots + n_p = n}} \binom{n}{n_1, n_2, \ldots, n_p} a_1^{n_1} a_2^{n_2} \cdots a_p^{n_p}$$

PROOF: Suppose $a_j^i \in A$ and form the product

$$(a_1^1 + a_2^1 + \cdots + a_p^1)(a_1^2 + a_2^2 + \cdots + a_p^2) \cdots (a_1^n + a_2^n + \cdots + a_p^n).$$

Given the nonnegative integers $n_1, n_2, \ldots, n_p$ ($n_1 + n_2 + \cdots + n_p = n$), consider each monomial of the form

$$(a_1^{i_1} a_1^{i_2} \cdots a_1^{i_{n_1}})(a_2^{j_1} a_2^{j_2} \cdots a_2^{j_{n_2}}) \cdots (a_p^{k_1} a_p^{k_2} \cdots a_p^{k_{n_p}});$$

each such monomial corresponds uniquely to an arrangement of the set $N = \{1, 2, \ldots, n\}$ into boxes $N_1, N_2, \ldots, N_p$, where $|N_1| = n_1$, $|N_2| = n_2, \ldots, |N_p| = n_p$, and conversely; the number of such monomials is therefore

$$\binom{n}{n_1, n_2, \ldots, n_p} = \frac{n!}{n_1! n_2! \cdots n_p!}.$$

The formula is obtained by putting

$$a_i^1 = a_i^2 = \cdots = a_i^n = a_i$$

for each i.

CONSEQUENCE 1

$$\binom{n}{n_1, n_2, \ldots, n_p} = \sum_{\substack{i \\ n_i \neq 0}} \binom{n-1}{n_1, n_2, \ldots, n_{i-1}, n_i - 1, n_{i+1}, \ldots, n_p}.$$

PROOF:

$$(a_1 + a_2 + \cdots + a_p)^n = (a_1 + a_2 + \cdots + a_p)(a_1 + a_2 + \cdots + a_p)^{n-1}.$$

The general term is

$$\binom{n}{n_1, n_2, \ldots, n_p} a_1^{r_i} a_2^{n_2} \cdots a_p^{n_p}$$

$$= \sum_{\substack{i \\ n_i \neq 0}} a_i \binom{n-1}{n_1, \ldots, n_i - 1, \ldots, n_p} a_1^{n_1} \cdots a_i^{n_i - 1} \cdots a_p^{n_p}.$$

CONSEQUENCE 2

$$\binom{m+q}{n_1, n_2, \ldots, n_p}$$

$$= \sum_{\substack{(k_1, k_2, \ldots) \leqslant (n_1, n_2, \ldots) \\ k_1 + k_2 + \cdots = m}} \binom{m}{k_1, k_2, \ldots, k_p} \binom{q}{n_1 - k_1, n_2 - k_2, \ldots, n_p - k_p}.$$

PROOF:

$$(a_1 + a_2 + \cdots + a_p)^{m+q} = (a_1 + a_2 + \cdots + a_p)^m (a_1 + a_2 + \cdots + a_p)^q.$$

The general term is

$$\binom{m+q}{n_1, n_2, \ldots, n_p} a_1^{n_1} a_2^{n_2} \cdots a_p^{n_p}$$

$$= \sum_{\substack{(k_1, k_2, \ldots) \leqslant (n_1, n_2, \ldots) \\ k_1 + k_2 + \cdots + k_p = m}} \binom{m}{k_1, k_2, \ldots, k_p} a_1^{k_1} a_2^{k_2} \cdots a_p^{k_p}$$

$$\times \binom{q}{n_1 - k_1, n_2 - k_2, \ldots, n_p - k_p} a_1^{n_1 - k_1} a_2^{n_2 - k_2} \cdots a_p^{n_p - k_p}.$$

(These formulas are analogous to Formulas 1 and 2 of Section 8, and, like them, they can be derived directly.)

CONSEQUENCE 3

$$\sum \binom{n}{n_1, n_2, \ldots, n_p} (-1)^{n_2 + n_4 + n_6 + \cdots} = \frac{1 - (-1)^p}{2}.$$

PROOF: Let

$$+1 = a_1 = a_3 = a_5 = \cdots,$$

$$-1 = a_2 = a_4 = a_6 = \cdots.$$

If p is even or odd,

$$(a_1 + a_2 + \cdots + a_p)^n = \frac{1 - (-1)^p}{2},$$

whence the formula.

Many other analogous formulas can be obtained by these methods.

APPLICATION TO THE LATTICE OF p-TUPLES.   Consider the set N^p of p-tuples $a = (a_1, a_2, \ldots, a_p)$, where $a_1, a_2, \ldots, a_p$ are nonnegative integers. Write

$$b \leqslant a$$

if $b_i \leqslant a_i$, $i = 1, 2, \ldots, p$.

The relation $\leqslant$ is an *order relation*, i.e.,

(1) $a \leqslant a$,
(2) $a \leqslant b, b \leqslant a$ implies $a = b$,
(3) $a \leqslant b, b \leqslant c$ implies $a \leqslant c$.

Write

$$a \vee b = (\max\{a_1, b_1\}, \max\{a_2, b_2\}, \ldots, \max\{a_p, b_p\}),$$

$$a \wedge b = (\min\{a_1, b_1\}, \min\{a_2, b_2\}, \ldots, \min\{a_p, b_p\}).$$

Clearly, the ordered set N^p is a lattice, that is, for any two p-tuples a and b, there exist a least upper bound, $a \vee b$, and a greatest lower bound, $a \wedge b$.

Let us represent this lattice by a graph in which the vertices are the p-tuples and two vertices a and b are joined by an edge directed from a to b if there exists an i such that

$$a_k = \begin{cases} b_k & \text{if } k \neq i, \\ b_i + 1 & \text{if } k = i. \end{cases}$$

If $a \geqslant b$, let us try to find the number of paths (following the direction of the arrows) going from a to b. For example, if $p = 3$, there are 10 paths from 032 to 000:

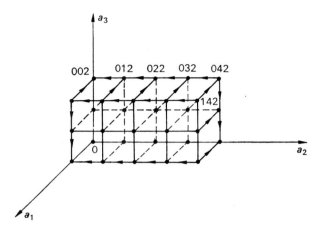

032   022, 012, 002, 001   000
     022, 012, 011, 001
     022, 012, 011, 010
     022, 021, 011, 001
     022, 021, 011, 010
     022, 021, 020, 010
     031, 021, 011, 001
     031, 021, 011, 010
     031, 021, 020, 010
     031, 030, 020, 010

PROPOSITION. *If $a \geqslant b$, the number of paths from a to b is*

$$\left(\begin{array}{c} \sum (a_i - b_i) \\ a_1 - b_1, a_2 - b_2, \ldots, a_n - b_n \end{array} \right).$$

PROOF: Let $a_i - b_i = n_i$, $\sum (a_i - b_i) = n$, and define the operation α_i by

$$\alpha_i(a) = (a_1, a_2, \ldots, a_i - 1, a_{i+1}, \ldots, a_n).$$

Now associate with each path from a to b the corresponding sequence of operations α_i, one term of the sequence for each step of the path. This establishes a bijection between the paths from a to b and the

n-tuples containing α_1, n_1 times, α_2, n_2 times, etc. Such an n-tuple $(x_1, x_2, x_3, \ldots, x_n)$ defines an arrangement of 1, 2, $\ldots$, n into boxes:

$$A_k = \{i/x_i = \alpha_k\}.$$

For example $\alpha_1 \; \alpha_1 \; \alpha_3 \; \alpha_2 \; \alpha_3 \; \alpha_3$ corresponds to the arrangement of 1, 2, $\ldots$, 6 into three boxes

$$A_1 = \{1, 2\}, \qquad A_2 = \{4\}, \qquad A_3 = \{3, 5, 6\};$$

this correspondence is obviously a bijection.

Therefore, the number of required paths is

$$\binom{n}{n_1, n_2, \ldots, n_p}.$$

10. STIRLING NUMBERS $S_n{}^m$, OR PARTITIONS OF n OBJECTS INTO m CLASSES

Let X be a set; subsets A_1, A_2, $\ldots$, A_p of X form a *partition of X* if

$$A_i \neq \varnothing,$$

$$i \neq j \qquad \text{implies} \quad A_i \cap A_j = \varnothing,$$

$$A_1 \cup A_2 \cup \cdots \cup A_p = X.$$

The sets A_i are the *classes* of the partition. Partitions of a set X arise from relations such as "a and b are contained in the same subset." More precisely they arise from relations $a \sim b$ satisfying the following axioms:

$$a \sim a \qquad \text{(reflexivity);}$$
$$a \sim b \Rightarrow b \sim a \qquad \text{(symmetry);}$$
$$a \sim b, \, b \sim a \Rightarrow a \sim c \qquad \text{(transitivity).}$$

We say $\sim$ is an *equivalence relation* and that the A_i's are the equivalence classes of this relation.

Consider a partition $\mathscr{A} = (A_1, A_2, \ldots, A_p)$ consisting of

$$\lambda_1 \quad \text{classes of cardinality} \quad 1,$$
$$\lambda_2 \quad \text{classes of cardinality} \quad 2,$$
$$\lambda_3 \quad \text{classes of cardinality} \quad 3,$$
$$\vdots$$
$$\lambda_k \quad \text{classes of cardinality} \quad k.$$

Then the partition is said to be *of type*

$$\underbrace{1 + 1 + \cdots + 1}_{\lambda_1} + \underbrace{2 + 2 + \cdots}_{\lambda_2} + \underbrace{k + k + \cdots + k}_{\lambda_k}.$$

For the sake of brevity, multiplicative notation is sometimes used instead of additive notation to denote the type of partition, viz.,

$$1^{\lambda_1} 2^{\lambda_2} \cdots k^{\lambda_k}.$$

Both notations will be used below.

For example, consider a collection of objects where

5   are red,
5   are green,
2   are blue,
2   are yellow,
2   are black,
1   is orange.

Then the colors define a partition of type $1 + 2 + 2 + 2 + 5 + 5$ (first notation) or $1 \cdot 2^3 \cdot 5^2$ (second notation).

REMARK: Sometimes one considers a decomposition of the set X into subsets $A_1, A_2, \ldots, A_p$, satisfying only the axioms

$$i \neq j \Rightarrow A_i \cap A_j = \varnothing,$$

$$\bigcup_{i=1}^{k} A_i = X.$$

If the A_i's can be empty, we say "X is decomposed into subsets" rather than "X is partitioned into classes." A "class" is necessarily nonempty.

Consider partitions $\mathscr{A} = (A_1, A_2, \ldots, A_p)$ and $\mathscr{B} = (B_1, B_2, \ldots, B_q)$ of X. Write $\mathscr{A} \prec \mathscr{B}$ if

$$A_i \cap B_j \neq \varnothing \Rightarrow A_i \subset B_j.$$

The relation $\prec$ is an order relation, i.e.,

$$\mathscr{A} \prec \mathscr{A}$$

$$\mathscr{A} \prec \mathscr{B}, \quad \mathscr{B} \prec \mathscr{A} \Rightarrow \mathscr{A} = \mathscr{B}$$

$$\mathscr{A} \prec \mathscr{B}, \quad \mathscr{B} \prec \mathscr{C} \Rightarrow \mathscr{A} \prec \mathscr{C}.$$

These conditions are easily verified.

The set of partitions of a set X, together with the relation $\prec$, is a lattice since, for any two partitions $\mathscr{A}$ and $\mathscr{B}$, there exist a least upper bound and a greatest lower bound.

EXAMPLE. $X = \{a, b, c, d\}$. We obtain the following lattice:

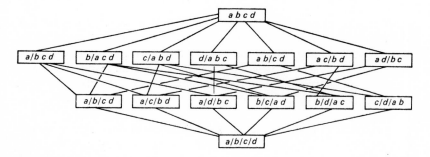

The types of the different partitions in this lattice are given by

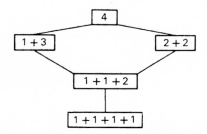

There is, therefore, a natural order relation on the types of partitions. However, the set of types of partitions together with this order relation does not form a lattice; for example the partitions $1 + 1 + 3$ and $1 + 2 + 2$ have two upper bounds $1 + 4$ and $2 + 3$, neither of which is "smaller" than the other.

PROPOSITION 1. *If X is a set of n objects, the number of partitions of type $1^{\lambda_1} 2^{\lambda_2} \cdots k^{\lambda_k}$ is*

$$\frac{n!}{(1!)^{\lambda_1}(2!)^{\lambda_2} \cdots (k!)^{\lambda_k}(\lambda_1!)(\lambda_2!)\cdots(\lambda_k!)}.$$

PROOF: This follows immediately from the proposition of Section 9.

EXAMPLE. $X = \{a, b, c, d\}$. From the figure above, we see that the number of partitions of type $1^2 2$ is

$$6 = \frac{4!}{(1!)^2(2!)^1(2!)}.$$

If $n \geqslant m$, the number S_n^m of partitions of a set of n objects into m classes is called the *Stirling number* (of *the second kind*). It is, for example, the number of distinct ways of arranging a set of n distinct objects into a collection of m identical boxes, not allowing any box to be empty (if empty boxes were permitted, then the number would be $S_n^1 + S_n^2 + \cdots + S_n^m$).

EXAMPLE. For four objects a, b, c, d, the number of partitions into two classes is 7, as can be seen from the above diagram. Therefore $S_4^2 = 7$.

PROPOSITION 2. *The number of surjections of X into A ($|A| = m$) is equal to $m! S_n^m$.*

PROOF: Each surjection of $X = \{1, 2, \ldots, n\}$ into $A = \{a_1, a_2, \ldots, a_m\}$ induces a partition of X into m distinct classes $a_1, a_2, \ldots, a_m$; conversely, each partition of X into m classes corresponds to $m!$ surjections of X into A.

RECURRENCE RELATIONS

$$S_{n+1}^k = S_n^{k-1} + kS_n^k, \qquad \text{where} \quad 1 < k < n,$$
$$S_n^1 = S_n^n = 1.$$

PROOF: Consider the table of partitions of $n + 1$ objects into k classes.

(1) For some of these partitions, the $(n + 1)$th object is the sole member of a class. The number of such partitions is S_n^{k-1}.

(2) For all other partitions, the $(n + 1)$th object is not the sole

member of any class. Hence, there are kS_n^k such partitions whence

$$S_{n+1}^k = S_n^{k-1} + kS_n^k.$$

These formulas enable us to calculate recursively the numbers S_n^k, thus:

S_n^m	$m = 1$	2	3	4	5	6
$n = 1$	1	0	0	0	0	0
$n = 2$	1	1	0	0	0	0
$n = 3$	1	3	1	0	0	0
$n = 4$	1	7	6	1	0	0
$n = 5$	1	15	25	10	1	0
$n = 6$	1	31	90	65	15	1

(handwritten margin notes: $S_n^{m-1} = \binom{n}{2}$; $S_n^2 = \binom{n}{1} + \cdots \binom{n-1}{2}$, n odd ; $= \binom{n}{1} + \cdots \frac{1}{2}\binom{n}{2}$, n even *)*

This is " Stirling's triangle."

From the triangles of Pascal (see Section 8) and Stirling the table of the number of mappings of X into A, which have as large a range of values in A as possible, is as follows:

	$m = 1$	2	3	4	5	6 ⋯
$n = 1$	1	2	3	4	5	6
2	1	2	6	12	20	30
3	1	6	6	24	60	120
4	1	14	36	24	120	360
5	1	30	150	240	120	720
6	1	62	540	1560	1800	720

$n < m$: The number of injections is $n!\binom{m}{n}$.

$n = m$: The number of bijections is $n!$.

$n > m$: The number of surjections is $m!\,S_n^m$.

FORMULA 1. *Let x be a real number. Then*

$$x^n = \sum_{k=1}^{n} S_n^k[x]_k$$

PROOF: Let $|A| = m \leqslant |X| = n$; the number of mappings of X into A is

$$m^n = \sum_{k=1}^{m} \binom{m}{k}(k!S_n^k) = \binom{m}{1}1!S_n^1 + \binom{m}{2}2!S_n^2 + \cdots + \binom{m}{n}n!S_n^n.$$

The above identity, of degree n, is therefore valid for $n + 1$ values of the variable x, viz., 0, 1, 2, ..., n. Hence, the identity is true for all x.

FORMULA 2

$$S_{n+1}^m = \sum_{k=0}^{n} \binom{n}{k}S_k^{m-1}$$

PROOF: Consider the table of partitions of $n + 1$ objects into m classes. In each partition, eliminate the class containing the $(n + 1)$th object; in this way we obtain all the partitions of a set K of objects into $m - 1$ classes [for each $K \subset \{1, 2, \ldots, n\}$], whence the formula.

EXERCISE. How many ways are there of placing n distinct objects into m different boxes in such a way that k of these boxes are nonempty and $m - k$ are empty? The answer is

$$(k!S_n^k)\binom{m}{k} = S_n^k[m]_k.$$

11. BELL EXPONENTIAL NUMBER B_n, OR THE NUMBER OF PARTITIONS OF n OBJECTS

Let B_n denote the number of partitions of a set X of n objects; these numbers are sometimes called *exponential numbers*, or *Bell numbers*.

EXAMPLE. $X = \{a, b, c, d\}$.
There are 15 partitions of X as can be seen from the figure in Section 10. Thus, $B_4 = 15$.

FORMULA 1

$$B_{n+1} = \sum_{k=0}^{n} \binom{n}{k} B_k$$

PROOF: Obviously, from the definition of Stirling numbers (Section 10) of the second kind,

$$B_n = S_n{}^1 + S_n{}^2 + \cdots + S_n{}^n.$$

If $m > n$, write $S_n{}^m = 0$; therefore, by Formula 2 of Section 10

$$B_{n+1} = \sum_{m=1}^{\infty} S_{n+1}^m = \sum_{m=1}^{\infty} \sum_{k=0}^{n} \binom{n}{k} S_k^{m-1} = \sum_{k=0}^{n} \binom{n}{k} \left(\sum_{m=1}^{\infty} S_k^{m-1} \right),$$

whence the formula.

FORMULA 2 (E. T. Bell)

$$\sum_{n=0}^{\infty} \frac{B_n}{n!} t^n = e^{(e^t - 1)}$$

PROOF: We give Rota's [7] elegant proof of this formula. Consider the vector space of real valued functions $\varphi(x)$ defined by

$$\varphi(x) = \sum_{n=0}^{\infty} \alpha_n [x]_n, \qquad \sum_{n=0}^{\infty} |\alpha_n| < + \infty.$$

The transformation $L(\varphi) = \sum_{n=0}^{\infty} \alpha_n$ is linear since

$$L(\lambda \varphi + \lambda' \varphi') = \sum_{n=0}^{\infty} (\lambda \alpha_n + \lambda' \alpha_n') = \lambda \sum_{n=0}^{\infty} \alpha_n + \lambda' \sum_{n=0}^{\infty} \alpha_n'.$$

On the other hand, from Formula 1 of Section 10,

$$L(x^n) = L\left(\sum_{k=1}^{n} S_n{}^k [x]_k \right) = \sum_{k=1}^{n} S_n{}^k = B_n.$$

Therefore

$$\sum_{n=0}^{\infty} \frac{B_n}{n!} t^n = \sum_{n=0}^{\infty} \frac{L(x^n)}{n!} t^n = L(e^{tx}).$$

Putting $e^t = (1 + u)$, we obtain

$$\sum_{n=0}^{\infty} \frac{B_n}{n!} t^n = L((1 + u)^x) = L\left(\sum_{n=0}^{\infty} \frac{[x]_n}{n!} u^n\right)$$

$$= \sum_{n=0}^{\infty} \frac{u^n}{n!} L([x]_n) = \sum_{n=0}^{\infty} \frac{u^n}{n!} = e^u = e^{(e^t - 1)}.$$

FORMULA 3 (G. Dobinski)

$$B_{n+1} = \frac{1}{e}\left(1^n + \frac{2^n}{1!} + \frac{3^n}{2!} + \frac{4^n}{3!} + \cdots\right)$$

PROOF:

$$e = \sum_{k=0}^{\infty} \frac{1}{k!} = \sum_{k=n}^{\infty} \frac{1}{(k-n)!} = \sum_{k=n}^{\infty} \frac{[k]_n}{k!} = \sum_{k=0}^{\infty} \frac{[k]_n}{k!}.$$

The transformation L, defined above, satisfies

$$L([x]_n) = 1 = \frac{1}{e} \sum_{k=0}^{\infty} \frac{[k]_n}{k!}.$$

If $\varphi(x) = \sum_n \alpha_n [x]_n$, then

$$L(\varphi(x)) = \sum_n \alpha_n \sum_k \frac{1}{e} \frac{[k]_n}{k!} = \frac{1}{e} \sum_k \frac{1}{k!} \sum_n \alpha_n [k]_n = \frac{1}{e} \sum_k \frac{\varphi(k)}{k!}.$$

Letting $\varphi(x) = x^{n+1}$, we obtain Formula 3.

REFERENCES

1. E. T. BELL, Exponential polynomials, *Ann. of Math. II* **35**, 258–77 (1934).
2. C. BERGE, "Théorie des Graphes et ses Applications." Dunod, Paris, 1958.
3. C. BERGE, Sur un noveau calcul symbolique et ses applications, *J. Math. Pures et Appl.* **29**, 245–274 (1950).
4. P. ERDÖS and G. SZEKERES, A Combinatorial problem in geometry, *Composito Math.* **2**, 463–470 (1935).
5. C. FRASNAY, Quelques problèmes combinatoires concernant les ordres totaux, doctoral thesis, Paris, 1965; *Ann. Inst. Fourier* **15**, 415–524 (1965).

6. J. RIORDAN, "Introduction to Combinatorial Analysis." Wiley, New York, 1958.
7. G. C. ROTA, The number of partitions of a set, *Amer. Math. Monthly*, **71**, 498–504 (1964).
8. H. J. RYSER, "Combinatorial Mathematics." Mathematical Assn. of America, Buffalo, 1963.
9. J. TOUCHARD, Nombres exponentials et nombre de Bernoulli, *Canad. J. Math.* **8**, 305–320 (1956).
10. A. KAUFMANN, "Initiation à la Combinatorique." Dunod, Paris, 1968.

CHAPTER 2

PARTITION PROBLEMS

1. P_n^m, OR THE NUMBER OF PARTITIONS OF INTEGER n INTO m PARTS

In Chapter 1 we counted the partitions of a set X into classes; we now intend to count the different *types* of partitions of a set of n objects. Each type corresponds to an m-tuple $(\alpha_1, \alpha_2, \ldots, \alpha_m)$ satisfying

$$n = \alpha_1 + \alpha_2 + \alpha_3 + \cdots + \alpha_m,$$

$$\alpha_1 \geqslant \alpha_2 \geqslant \alpha_3 \geqslant \cdots \geqslant \alpha_m \geqslant 1.$$

This m-tuple is usually called a *partition of the integer n*. The number of partitions of the integer n into exactly m classes (called "*parts*" in this context) is denoted by P_n^m.

EXAMPLE

The partitions of	are	whence
2	2 and 1 + 1	$P_2^{\,1} = P_2^{\,2} = 1$
3	3 and 2 + 1 and 1 + 1 + 1	$P_3^{\,1} = P_3^{\,2} = P_3^{\,3} = 1$
4	4 and 3 + 1 and 2 + 2 and 2 + 1 + 1 and 1 + 1 + 1 + 1	$P_4^{\,2} = 2$ $P_4^{\,1} = P_4^{\,3} = P_4^{\,4} = 1$

RECURRENCE RELATIONS. We have

$$P_n^{\ 1} + P_n^{\ 2} + \cdots + P_n^{\ k} = P_{n+k}^{k},$$
$$P_n^{\ 1} = P_n^{\ n} = 1.$$

PROOF: The second formula follows immediately from the definitions. We therefore need prove only the first formula. Let E be the set of partitions of n having k or less parts; each partition belonging to E may be considered as a k-tuple.

Define on E the mapping

$$(\alpha_1, \alpha_2, \ldots, \alpha_m, 0, 0, \ldots, 0) \rightarrow (\alpha_1 + 1, \alpha_2 + 1, \ldots, \alpha_m + 1, 1, 1, \ldots, 1).$$

E is mapped into the set E' of partitions of $n + k$ into exactly k parts. This mapping is bijective, since

(1) two distinct k-tuples of E are mapped onto two distinct k-tuples of E';

(2) every k-tuple of E' is the image of a k-tuple of E.

Therefore,

$$|E| = P_n^{\ 1} + \cdots + P_n^{\ k} = |E'| = P_{n+k}^{k}.$$

From these formulas the $P_n^{\ m}$'s may be calculated recursively, e.g.,

$P_n^{\ m}$	$m=1$	2	3	4	5	6	$\cdots$
$n = 1$	1	0	0	0	0	0	
2	1	1	0	0	0	0	
3	1	1	1	0	0	0	
4	1	2	1	1	0	0	
5	1	2	2	1	1	0	
6	1	3	3	2	1	1	

PROPOSITION 1. *The number of partitions of n into k parts is equal to the number of partitions of n into parts the largest of which is k.*

EXAMPLE. When $n = 6$, there are three partitions having 3 as the largest part: $3 + 1 + 1 + 1, 3 + 2 + 1, 3 + 3$; there are three partitions into three parts: $4 + 1 + 1, 3 + 2 + 1, 2 + 2 + 2$.

PROOF OF PROPOSITION: Consider a partition, for example, $5 + 4 + 1 + 1$, and associate with it a diagram (called the *Ferrer's diagram*) in which each part is represented by a row of squares equal in number to the part itself, and arranged as follows:

$$5 + 4 + 1 + 1 \quad \rightarrow$$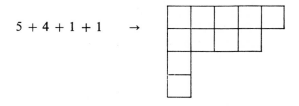

We call $\alpha^* = (\alpha_1^*, \alpha_2^*, \ldots)$ the *conjugate partition* of the partition $\alpha = (\alpha_1, \alpha_2, \ldots, \alpha_k)$, where α_i^* is the number of parts of α of cardinality i or more. The Ferrer's diagram associated with this partition is obtained from the original diagram by a rotation about the diagonal.

The partition conjugate to $5 + 4 + 1 + 1$ is $4 + 2 + 2 + 2 + 1$, as can be deduced immediately from the Ferrer's diagram above: we compute the number of squares in the first column, then the number of squares in the second column, etc.

The mapping from the set of partitions of n into the set of conjugate partitions is bijective; we have therefore established a bijection between the set of partitions of n having k as the largest part and the set of partitions of n into k parts.

PROPOSITION 2. *The number of self-conjugate partitions of n is the same as the number of partitions of n with all parts unequal and odd.*

PROOF: Consider the Ferrer's diagram associated with a partition of n with all parts unequal and odd. We may obtain a new Ferrer's diagram by placing the squares of each row in a "set-square arrangement" as indicated below:

$9 + 3 + 1 \quad \rightarrow$ 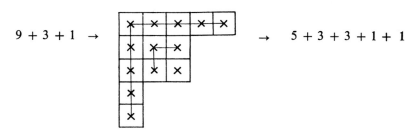 $\rightarrow \quad 5 + 3 + 3 + 1 + 1$

This Ferrer's diagram defines a self-conjugate partition (in this case, $5 + 3 + 3 + 1 + 1$). Similarly, by reversing the above construction, each self-conjugate partition corresponds to a unique partition with all parts unequal and odd.

PROPOSITION 3. *The number of partitions of n into unequal parts is equal to the number of partitions of n into odd parts.*

PROOF: Consider a partition into odd parts, for example,

$$\underbrace{5 + 5 + 5 + 5}_{4} + \underbrace{3 + 3 + 3}_{3} + \underbrace{1 + 1 + 1 + 1 + 1}_{5}.$$

Each exponent can be expressed uniquely as a sum of powers of 2 (binary decomposition); in this case,

$$4 = 2^2,$$
$$3 = 2^0 + 2,$$
$$5 = 2^0 + 2^2.$$

Then the 2^k identical parts in the Ferrer's diagram may be grouped together as below:

We group together the squares in the same bracket.

Since each number can be written uniquely as the product of an odd number and a power of 2, this defines a new partition $20 + 6 + 4 + 3 + 1$, all parts of which are unequal. This correspondence between partitions of n into odd parts and partitions of n into unequal parts is clearly a bijection. This completes the proof.

PROPOSITION 4. *Let Q_n be the number of partitions of n into unequal parts and with an even number of parts; let $Q_n{'}$ be the number of partitions of n into unequal parts and with an odd number of parts. Then*

$$Q_n = \begin{cases} Q_n{'} & \text{if } n \neq (k/2)(3k \pm 1), \\ Q_n{'} + (-1)^k & \text{if } n = (k/2)(3k \pm 1). \end{cases}$$

PROOF: Consider a partition of n into an odd number of parts, no two of which are equal: in the Ferrer's diagram of this partition let S

be the set of squares in the horizontal line to the extreme south; and let
E be the set of squares in the line at 45° to the extreme east (these two
lines may consist of just one unique square).

For example,

$$7 + 6 + 5 + 3 + 2 \quad \rightarrow$$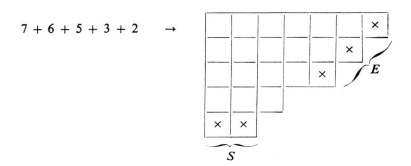

If $|S| \leqslant |E|$, move the line S to the extreme east of the Ferrer's
diagram, thus:

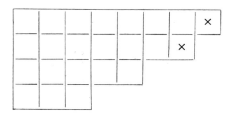

Hence, we obtain a new partition $8 + 7 + 5 + 3$ with an even
number of parts no two of which are equal (and with $|S| > |E|$). If
initially $|S| > |E|$, move the line E to the extreme south so as to obtain
a new partition into an even number of parts no two of which are equal
(and with $|S| \leqslant |E|$).

These operations are always possible except when the lines S and E
both meet, and, in addition,

$$|S| = |E|, \quad \text{or} \quad |S| = |E| + 1.$$

CASE $|S| = |E|$: S cannot be moved to the east.

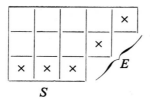

CASE $|S| = |E| + 1$: E cannot be moved to the south.

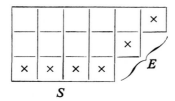

In the exceptional cases, let $|E| = k$, $\varepsilon = 0$ or 1:

$$n = (k + \varepsilon) + (k + \varepsilon + 1) + (k + \varepsilon + 2) + \cdots + (2k + \varepsilon - 1)$$

$$= \frac{k}{2}(3k + 2\varepsilon - 1)$$

$$= \frac{k}{2}(3k \pm 1).$$

In all the other cases the correspondence we have established above is bijective; hence the above formulas. **Q.E.D.**

Another way of studying the numbers $P_n{}^m$ is to use the *method of generating functions* due to Laplace and Euler.

Let $\varphi(n)$ be a function defined on the integers $n \geqslant 0$; associate with it the function

$$F(x) = \sum_{n=0}^{\infty} \varphi(n)x^n$$

(we will always assume that this series converges in the neighborhood of the origin). $F(x)$ is said to be the *generating function* of $\varphi(n)$, and we write

$$\varphi(n) \doteqdot F(x).$$

$\varphi(n)$ is said to be the *image* of the function $F(x)$, and it follows that

$$\varphi(n) = \frac{1}{n!} \left[\frac{d^n F}{dx^n} \right]_0.$$

(1) The generating function of the number P_n of partitions of the integer n is

$$F_1(x) = (1 - x)^{-1}(1 - x^2)^{-1}(1 - x^3)^{-1} \cdots$$

(assuming that $P_0 = 1$).

PROOF: We must prove that

$$\frac{1}{(1 - x)(1 - x^2)(1 - x^3) \cdots} = \sum_{n=0}^{\infty} P_n x^n.$$

Now, when $|x| < 1$,

$$\frac{1}{(1 - a_1 x)(1 - a_2 x^2) \cdots (1 - a_k x^k) \cdots}$$

$$= (1 + a_1 x + a_1{}^2 x^2 + a_1{}^3 x^3 + \cdots)$$

$$\times (1 + a_2 x^2 + a_2{}^2 x^4 + \cdots) \cdots (1 + a_k x^k + a_k{}^2 x^{2k} + \cdots) \cdots$$

$$= 1 + a_1 x + \cdots + (a_1^{\lambda_1} a_2^{\lambda_2} \cdots a_k^{\lambda_k} + \cdots)x^n + \cdots.$$

In the coefficient of x^n each term $a_1^{\lambda_1} a_2^{\lambda_2} \cdots a_k^{\lambda_k}$ defines a partition of n, viz.,

$$\underbrace{(1 + 1 + \cdots + 1)}_{\lambda_1} + \underbrace{(2 + 2 + \cdots + 2)}_{\lambda_2} + \cdots + \underbrace{(k + k + \cdots + k)}_{\lambda_k} = n.$$

In this way all the partitions of n are obtained with neither repetitions nor omissions. Finally, putting $1 = a_1 = a_2 = \cdots$, we obtain the above formula.

(2) The generating function of $P_n{}^m$ is

$$F_2(x) = x^m(1 - x)^{-1}(1 - x^2)^{-1} \cdots (1 - x^m)^{-1}.$$

(3) The generating function of the number of partitions of n consisting only of odd parts is

$$F_3(x) = (1 + x + x^2 + \cdots)(1 + x^3 + x^6 + \cdots)(1 + x^5 + x^{10} + \cdots) \cdots$$
$$= (1 - x)^{-1}(1 - x^3)^{-1}(1 - x^5)^{-1} \cdots.$$

(4) The generating function of the number of partitions of n into unequal parts is

$$F_4(x) = (1 + x)(1 + x^2)(1 + x^3) \cdots.$$

(5) The generating function of the number of partitions of n into odd parts no two of which are equal is

$$F_5(x) = \prod_{i=0}^{\infty} (1 + x^{2i+1}).$$

APPLICATION OF CERTAIN IDENTITIES TO THE COUNTING OF PARTITIONS. Recall, for example, Proposition 3: the number of partitions of n into unequal parts is equal to the number of partitions of n into odd parts.

This is simply a question of showing that

$$F_3(x) = \frac{1}{(1 - x)(1 - x^3)(1 - x^5) \cdots}$$

and

$$F_4(x) = (1 + x)(1 + x^2)(1 + x^3) \cdots$$

are identical.
Indeed, one has

$$F_3(x) = \frac{(1 - x^2)(1 - x^4)(1 - x^6) \cdots}{(1 - x)(1 - x^2)(1 - x^3) \cdots}$$
$$= \frac{(1 - x)(1 + x)(1 - x^2)(1 + x^2) \cdots}{(1 - x)(1 - x^2) \cdots}$$
$$= F_4(x).$$

This proof, due to Euler, is very general and does not involve any exceptional intuition; on the debit side however, it does not yield explicitly a bijection between the two classes of partitions.

APPLICATION OF THE COUNTING OF PARTITIONS TO CERTAIN IDENTITIES. Consider Euler's identity

$$(1 - x)(1 - x^2)(1 - x^3) \cdots = 1 + \sum \varphi(n)x^n = 1 - x - x^2 + x^5 + \cdots.$$

The image of this generating function is

$$\varphi(n) = \begin{cases} 0 & \text{if} \quad n \neq \dfrac{3\,k^2 \pm k}{2}, \\[3mm] (-1)^k & \text{if} \quad n = \dfrac{3\,k^2 \pm k}{2}. \end{cases}$$

PROOF: Let Q_n be the number of partitions of n into unequal parts, and with an even number of parts; let Q_n' be the number of partitions of n into unequal parts, and with an odd number of parts.

The coefficient of x^n, in the product above, is clearly equal to $Q_n - Q_n'$; by Proposition 4, this number is equal to $\varphi(n)$. **Q.E.D.**

Other identities exist of this type, the most famous being those of Jacobi and Rogers–Ramanujan; remarkably they can all be interpreted in terms of partitions. In addition, they enable us to study the asymptotic behavior of P_n and P_n^m.

Hence, Hardy and Ramanujan proved

$$\log P_n \sim \pi(2n/3)^{1/2} - \log[4n(3)^{1/2}]$$

(see Wright [12]).

2. $P_{n,h}$, OR THE NUMBER OF PARTITIONS OF THE INTEGER n HAVING h AS THE SMALLEST PART

Let

P_n denote the number of partitions of n,
P_n^m denote the number of partitions of n into m parts,

$P_{n,h}$ denote the number of partitions of n having h as the
            smallest part,
$P_{n,h}^m$ denote the number of partitions of n into m parts having
            h as the smallest part.

Then obviously

$$P_n = P_n{}^1 + P_n{}^2 + \cdots + P_n{}^n,$$

$$P_n{}^m = P_{n,1}^m + P_{n,2}^m + \cdots + P_{n,n}^m,$$

$$P_{n,h} = P_{n,h}^1 + P_{n,h}^2 + \cdots + P_{n,h}^n.$$

EXAMPLE. Consider the partitions of 6 into m parts having h as the
smallest part. These partitions may be represented by a tree, as follows:

Level:

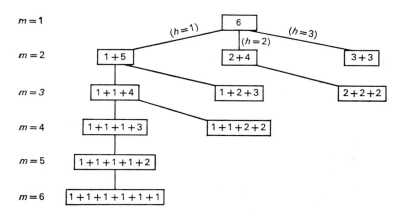

This representation is extremely convenient for finding the number
$P_{n,h}^m$ quickly.   Looking down the tree, one sees immediately that
$P_{6,1}^4 = 2$, etc. (Bertier [2]).

FORMULA 1

$$P_{n,h}^m = P_{n-m,h-1}^m \qquad \text{if} \quad h > 1,$$
$$P_{n,h}^m = P_{n-1}^{m-1} \qquad \text{if} \quad h = 1.$$

PROOF: Let us suppose $h > 1$; each partition $n = \alpha_1 + \alpha_2 + \cdots + \alpha_m$, with

$$\alpha_1 \geqslant \alpha_2 \geqslant \cdots \geqslant \alpha_m = h,$$

can be associated with the partition

$$n - m = (\alpha_1 - 1) + (\alpha_2 - 1) + \cdots + (\alpha_{m-1} - 1) + (h - 1).$$

FORMULA 2

$$P_{n,h} = P_{n-h,h} + P_{n-h,h+1} + \cdots$$

PROOF: Since

$$n = h + (n - h),$$

$$P_{n,h} = \sum_{k \geqslant h} P_{n-h,k}.$$

RECURRENCE RELATIONS

$$P_{n,h} = P_{n-1,h-1} - P_{n-h,h-1} \qquad \text{if} \quad h > 1,$$
$$P_{n,h} = P_{n-1} \qquad\qquad\qquad\qquad \text{if} \quad h = 1.$$

PROOF: Let us suppose $h > 1$; from Formula 2,

$$P_{n-1,h-1} = \sum_{k \geqslant h-1} P_{n-h,k}.$$

Comparing this formula with Formula 2, we obtain

$$P_{n,h} = P_{n-1,h-1} - P_{n-h,h-1}.$$

Using this formula, column by column, the following table can be constructed:

$P_{n,h}$	$h=1$	$h=2$	$h=3$	$h=4$	$h=5$ $\cdots$	P_n
$n=1$	1	0	0	0	0 $\cdots$	1
$n=2$	1	1	0	0	0 $\cdots$	2
$n=3$	2	0	1	0	0 $\cdots$	3
$n=4$	3	1	0	1	0 $\cdots$	5
$n=5$	5	1	0	0	1 $\cdots$	7
$n=6$	7	2	1	0	0 1 $\cdots$	11
$n=7$	11	2	1	0	0 $\cdots$ 1 $\cdots$	15
$n=8$	15	4	1	1	0 $\cdots$ 1 $\cdots$	22
$n=9$	22	4	2	1	0 $\cdots$ 1 $\cdots$	30
$n=10$	30	7	2	1	1 $\cdots$ 1 $\cdots$	42
$\vdots$						

3. COUNTING THE STANDARD TABLEAUS ASSOCIATED WITH A PARTITION OF n

Given a partition, $\alpha_1 + \alpha_2 + \cdots + \alpha_m = n$ $(\alpha_1 \geq \alpha_2 \geq \cdots \geq \alpha_m)$ of n, an array $((k_j{}^i))$ of integers, such that the ith row $(k_1{}^i, k_2{}^i, k_3{}^i, \ldots, k_{\alpha_i}^i)$ of the array contains exactly α_i integers, is called a *tableau associated with this partition*.

The tableau is *normal* if

(1) the $k_j{}^i$'s are all distinct;
(2) $i > j \Rightarrow k_s{}^i > k_s{}^j$;
(3) $i > j \Rightarrow k_i{}^t > k_j{}^t$.

For example, the array

$$((k_j{}^i)) = \begin{array}{|c|c|c|} \hline 1 & 3 & 6 \\ \hline 2 & 4 \\ \cline{1-2} 5 & 7 \\ \cline{1-2} \end{array}$$

is a normal tableau associated with the partition $3 + 2 + 2$: notice that the integers increase reading from left to right of a row, or reading from top to bottom of a column.

A normal tableau associated with a partition of n, in which only the integers 1, 2, ..., n are used, is called a *standard tableau*. The *order* of this standard tableau is n. Our aim now is to count the number of standard tableaus associated with a given partition; this problem has applications in the theory of the representations of the symmetric group.*

EXAMPLE. The number of standard tableaus associated with the partition $3 + 2 + 2$ is 21; the list is as follows:

. . .	123	124	134	125	135	
. .	45	35	25	34	24	
67	67	67	67	67	67	
. . .	123	124	134	125	135	145
. 6	46	36	26	36	26	26
. 7	57	57	57	47	47	37
. . 6	126	136	126	136	146	
. .	34	24	35	25	25	
. 7	57	57	47	47	37	
. . 7	127	137	127	137	147	
. .	34	24	35	25	25	
. 6	56	56	46	46	36	

Notice that the largest integer of this tableau (in this case, 7) must necessarily appear at the extreme right of a row, and at the bottom of a column (in this case, there are only two possibilities). Starting with the last row, all the possible places for 7 are tried successively.

Having fixed the position of 7, 6 is placed successively in all the possible positions at the extreme right of a row, and at the bottom of a column in the new diagram (in this case, there are three possible places), etc. . . .

* See Rutherford [9].

Given a partition

$$\alpha_1 + \alpha_2 + \cdots + \alpha_m = n, \qquad \alpha_1 \geqslant \alpha_2 \geqslant \cdots \geqslant \alpha_m \geqslant 1,$$

the array of numbers

$$h_j{}^i = 1 + (\alpha_i - j) + (\alpha_j{}^* - i),$$

is called a *square tableau*. This tableau may be obtained by representing the partition by its Ferrer's diagram.

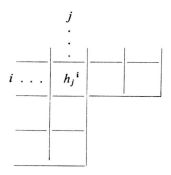

In the (i, j)th box, at the intersection of the ith row and jth column, we place the integer obtained by counting the box (i, j), then the boxes to its right (numbering $\alpha_i - j$), then the boxes below it (numbering $a_j{}^* - i$), and summing the result. Thus, the square tableau, corresponding to the partition $3 + 2 + 2$, is

$$
((h_j{}^i)) =
\begin{array}{|c|c|c|}
\hline
5 & 4 & 1 \\
\hline
3 & 2 \\
\cline{1-2}
2 & 1 \\
\cline{1-2}
\end{array}
$$

LEMMA: *Given a box (i, j) the sequence*

$$(h_j{}^i, h_{j+1}^i, \ldots, h_{\alpha_i}^i, h_j{}^i - h_j^{i+1}, h_j{}^i - h_j^{i+2}, \ldots, h_j{}^i - h_j^{\alpha_j*})$$

is a permutation of the integers $1, 2, \ldots, h_j^i$.

PROOF: Consider the "set square" of boxes, at the angle of which is the box (i, j),

i, j	$i, j + 1$	$i, j + 2$	$\cdots$	$i, j + p$
$i + 1, j$				
$i + 2, j$				
$\vdots$				
$i + q, j$				

where $p = (\alpha_i - j)$ and $q = (\alpha_j^* - i)$. This set square contains h_j^i boxes.

In the box (i, j) place the number h_j^i; in box $(i, j + 1)$ place the number h_{j+1}^i etc.; in box $(i + 1, j)$ place the number $h_j^i - h_j^{i+1}$ etc.

We show no number is inserted more than once; the sequence of entries in the first row of the set square is obviously strictly decreasing; the sequence of entries below (i, j) is obviously strictly increasing with last term

$$h_j^i - h_j^{\alpha_j^*} < h_j^i.$$

It therefore suffices to show that for $t \geqslant 1$, $s \geqslant 1$, we cannot have

$$h_{j+t}^i = h_j^i - h_j^{i+s}.$$

Consider three set squares at the angles of which are, respectively, the boxes (i, j), $(i, j + t)$, and $(i + s, j)$; illustrated by

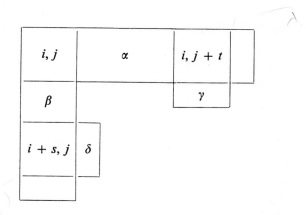

where α is the number of boxes between (i, j) and $(i, j + t)$, β is the number of boxes between (i, j) and $(i + s, j)$, γ is the number of boxes below $(i, j + t)$, and δ is the number of boxes to the right of $(i + s, j)$.

The above equality may be written

$$h^i_{j+t} + h^{i+s}_j = h^i_j,$$

or

$$\gamma + \delta = \alpha + \beta + 1.$$

If $\delta \leqslant \alpha$, then $\gamma \leqslant \beta$, and this equality gives $\gamma + \delta \geqslant \gamma + \delta + 1$, which is absurd. If $\delta > \alpha$, the set squares with angles $(i + s, j)$ and $(i, j + t)$ intersect and therefore $\gamma > \beta$ and the equality gives $\gamma + \delta = \alpha + \beta + 1 \leqslant (\delta - 1) + (\gamma - 1) + 1 = \gamma + \delta - 1$, which again is absurd. **Q.E.D.**

THEOREM (Frame–Robinson–Thrall [3]). *Let $((h_j{}^i))$ be the square tableau corresponding to a given partition*

$$\alpha_1 + \alpha_2 + \cdots + \alpha_m = n$$

of the integer n. The number of standard tableaus associated with the given partition is

$$f(\alpha_1, \ldots, \alpha_m) = n! / \prod_{i, j} h_j{}^i.$$

PROOF: By the lemma,

$$\prod_{i,j} h_j{}^i = \prod_{i=1}^{m} \prod_{j=1}^{\alpha_i} h_j{}^i = \frac{\prod_{i \geqslant 1} (h_1{}^i)!}{\prod_{\substack{i \geqslant 1 \\ j > i}} (h_1{}^i - h_1{}^j)}.$$

Write

$$x_i = h_1{}^i = \alpha_i + (m - i).$$

The function f, defined above, can then be written

$$f(\alpha_1, \alpha_2, \ldots, \alpha_m) = n! \; \frac{\prod_{\substack{i \geqslant 1 \\ j > i}} (x_i - x_j)}{\prod_{i \geqslant 1} (x_i)!}.$$

Furthermore,

$$f(\alpha_1, \alpha_2, \ldots, \alpha_{k-1}, \alpha_k - 1, \alpha_{k+1}, \ldots, \alpha_m)$$

$$= (n-1)! \; \frac{\prod_{\substack{i,j \neq k \\ j > i}} (x_i - x_j) \prod_{j > k} (x_k - 1 - x_j) \prod_{i < k} (x_i - x_k + 1)}{\prod_{i \neq k} (x_i)! \, (x_k - 1)!}$$

$$= \frac{x_k}{n} f(\alpha_1, \alpha_2, \ldots, \alpha_m) \prod_{j \neq k} \frac{(x_k - 1 - x_j)}{(x_k - x_j)}.$$

Notice that if $\alpha_{k+1} = \alpha_k$, then $x_{k+1} = x_k - 1$, and this expression vanishes.

Proceed by induction: the theorem is obviously valid for all partitions of $n = 1$ and $n = 2$; suppose the theorem is valid for partitions of integers less than n. We will prove that this implies the theorem is valid for the partition $\alpha_1 + \alpha_2 + \cdots + \alpha_m = n$.

In a standard tableau T_n, of order n, associated with the partition $\alpha_1, \alpha_2, \alpha_3, \ldots, \alpha_m$, n is placed at the end of the kth row (say), where $\alpha_{k+1} < \alpha_k$. By deleting from T_n the box containing the integer n, we obtain a standard tableau T_{n-1} associated with the partition $\alpha_1, \alpha_2, \ldots, \alpha_{k-1}, \alpha_k - 1, \alpha_{k+1}, \ldots, \alpha_m$ of $n - 1$, and hence a bijection is established between the set of all standard tableaus associated with the partition $\alpha_1, \alpha_2, \ldots, \alpha_m$, and a set of standard tableaus of order $n - 1$. Therefore, the number of standard tableaus of order n associated with the partition $\alpha_1, \alpha_2, \ldots, \alpha_m$ is

$$\sum_{k / \alpha_{k+1} \neq \alpha_k} f(\alpha_1, \ldots, \alpha_{k-1}, \alpha_k - 1, \alpha_{k+1}, \ldots, \alpha_m)$$

$$= \sum_{k=1}^{m} f(\alpha_1, \ldots, \alpha_{k-1}, \alpha_k - 1, \alpha_{k+1}, \ldots, \alpha_m)$$

$$= \frac{1}{n} f(\alpha_1, \ldots, \alpha_m) \sum_{k=1}^{m} x_k \prod_{j \neq k} \frac{(x_k - x_j - 1)}{(x_k - x_j)}.$$

Write

$$g(x) = \prod_{j=1}^{m} (x - x_j).$$

Now it is simply a question of verifying that if the m roots of a polynomial $g(x)$ satisfy

$$x_1 < x_2 < \cdots < x_m,$$

$$\sum_{i=1}^{m} x_i = \sum_{i=1}^{m} \alpha_i + \sum_{i=1}^{m} (m - i) = n + \frac{m(m-1)}{2},$$

then they also satisfy

$$n = \sum_{k=1}^{m} x_k \prod_{j \neq k} \frac{(x_k - x_j - 1)}{(x_k - x_j)} = \sum_{k=1}^{m} \frac{(-x_k)g(x_k - 1)}{g'(x_k)},$$

where

$$g'(x_k) = \frac{g(x)}{x - x_k}$$

(this verification, which is left to the reader, is a routine procedure in the theory of symmetric functions of the roots of a polynomial).

Robinson [8] gives* a remarkable construction, whereby any sequence φ of distinct integers is associated injectively with a pair $P(\varphi)$, $Q(\varphi)$ of normal tableaus with the same diagram. We now give this construction, which was presented independently by Schensted [10], whose terminology and notation we use.

Let T be a normal tableau formed with distinct integers $k_1, k_2, \ldots, k_n$. If

$$x \neq k_1, k_2, \ldots, k_n,$$

a new tableau can be formed using the following rules:

---

* We are grateful to M.P. Schützenberger [11] for this reference.

(1) If possible, insert x in the first row by replacing the smallest k_i in the first row which is greater than x by x. If no such k_i exists, add x to the end of the first row.

(2) If x replaces k_i in the first row insert k_i, following rule (1), in the second row, etc.

This new tableau is denoted by $(T \leftarrow x)$.

EXAMPLE. If

$$(T) = \begin{matrix} 2 & 4 & 7 \\ 3 & 8 \\ 5 & 9 \end{matrix} \, ,$$

then

$$(T \leftarrow 6) = \begin{matrix} 2 & 4 & 6 \\ 3 & 7 \\ 5 & 8 \\ 9 \end{matrix} \, .$$

Notice that $(T \leftarrow x)$ is again a normal tableau.

Given a sequence φ of distinct integers $k_1, k_2, \ldots, k_n$, let $P(\varphi)$ denote the normal tableau

$$[((k_1) \leftarrow k_2) \leftarrow k_3] \leftarrow k_4 \cdots .$$

Denote by $Q(\varphi)$ the tableau with the same diagram as $P(\varphi)$, obtained by inserting the integer i into the box which is added to the diagram when the integer k_i is first inserted in $P(\varphi)$.

EXAMPLE. Consider the sequence 3, 5, 4, 9, 8 and successively construct

$\varphi =$	3	3 5	3 5 4	3 5 4 9	3 5 4 9 8
$P(\varphi) =$	3	3 5	3 4	3 4 9	3 4 8
			5	5	5 9
$Q(\varphi) =$	1	1 2	1 2	1 2 4	1 2 4
			3	3	3 5

Conversely, if

$$P(\varphi) = \begin{matrix} 3 \ 4 \ 8 \\ 5 \ 9 \end{matrix} \quad \text{and} \quad Q(\varphi) = \begin{matrix} 1 \ 2 \ 4 \\ 3 \ 5 \end{matrix}$$

are given, the initial sequence φ can be easily reconstructed by reversing the above procedure. The last integer displaced in $P(\varphi)$ is 9, which can only have come from the first row when 8 displaced 9. Therefore, the last integer of the sequence is 8; etc.

Notice that $Q(\varphi)$ is a standard tableau; indeed, from the remark above, $P(\varphi)$ is a normal tableau and $Q(\varphi)$ has the same diagram. As each integer added to the tableau $Q(\varphi)$ is greater than those above it and to its left, $Q(\varphi)$ is again a normal tableau; therefore, it is a standard tableau.

LEMMA (Robinson [8]). *The number of columns of $P(\varphi)$ is equal to the length of the longest increasing subsequence of φ; the number of rows of $P(\varphi)$ is equal to the length of the longest decreasing subsequence of φ.*

PROOF:

(1) The sequence of distinct integers inserted into the jth box of the first row of the tableau $P(\varphi)$ is called the jth *fundamental sequence* of the sequence φ; obviously *every fundamental sequence is a decreasing subsequence of the initial sequence.*

The fundamental sequences in the preceding example are: 3; 5, 4; 9, 8.

(2) Given an integer k in the jth fundamental sequence, *there exists an integer, in the $(j-1)$th fundamental sequence, which is smaller than k and appears to the left of k in φ,* viz., the integer which is in the $(j-1)$th box of the first row of $P(\varphi)$ when k is inserted in the jth box of the first row of $P(\varphi)$.

(3) The number of columns of $P(\varphi)$ is equal to the number of fundamental sequences; from (1), there cannot exist more than one element from each fundamental sequence in an increasing subsequence of φ; from (2), an increasing subsequence of φ can be constructed with an element taken from each of the fundamental sequences. Therefore, the *number of columns of $P(\varphi)$ is the same as the length of the longest increasing subsequence of φ.*

(4) The second part of the lemma follows from observing that, on reversing the order of the sequence φ, the increasing subsequences are transformed into decreasing subsequences.

The argument above gives an easy method of constructing the subsequences of maximal length, viz., in the preceding example:

$$3, 4, 8 \quad \text{and} \quad 3, 4, 9.$$

THEOREM (Schensted [10]). *The number of permutations of* 1, 2, ..., *n that have a longest increasing subsequence of length p and a longest decreasing subsequence of length q is obtained by counting the standard tableaus associated with the partitions of n into q parts, the greatest part being equal to p, and summing the squares of the numbers obtained.*

PROOF: Corresponding to each permutation φ, satisfying the conditions of the theorem, are the two standard tableaus $P(\varphi)$ and $Q(\varphi)$. Both $P(\varphi)$ and $Q(\varphi)$ have the same diagram which, by the lemma, contains p columns and q rows; conversely, to every pair of standard tableaus of order n with the same diagram and containing p columns and q rows, there corresponds one, and only one, permutation satisfying the conditions of the theorem. The theorem follows.

EXAMPLE. For $n = 10$, $p = 6$, $q = 3$, consider the square tableaus

8	6	5	3	2	1
4	2	1			
1					

8	7	4	3	2	1
3	2				
2	1				

From Schensted's theorem, the required number of permutations is

$$\left(\frac{10!}{2 \cdot 3 \cdot 5 \cdot 6 \cdot 8 \cdot 4 \cdot 2}\right)^2 + \left(\frac{10!}{2 \cdot 3 \cdot 4 \cdot 7 \cdot 8 \cdot 2 \cdot 3 \cdot 2}\right)^2$$
$$= (315)^2 + (225)^2 = 149{,}850.$$

4. STANDARD TABLEAUS AND YOUNG'S LATTICE

Consider two sequences $\alpha = (\alpha_1, \alpha_2, \ldots, \alpha_p, 0, 0, \ldots)$, and $\beta = (\beta_1, \beta_2, \ldots, \beta_q, 0, 0, \ldots)$, with

$$\alpha_1 \geqslant \alpha_2 \geqslant \alpha_3 \geqslant \cdots \geqslant \alpha_p,$$
$$\beta_1 \geqslant \beta_2 \geqslant \beta_3 \geqslant \cdots \geqslant \beta_q.$$

These sequences correspond to partitions of the integers $\sum_i \alpha_i$ and $\sum_j \beta_j$, respectively. Write

$$\alpha \vee \beta = (\max\{\alpha_1, \beta_1\}, \max\{\alpha_2, \beta_2\}, \ldots),$$
$$\alpha \wedge \beta = (\min\{\alpha_1, \beta_1\}, \min\{\alpha_2, \beta_2\}, \ldots).$$

Then $\alpha \vee \beta$ and $\alpha \wedge \beta$ are decreasing sequences terminating in zeros. These operations define a lattice, called Young's lattice. The "bottom" of this lattice is as follows:

Level:

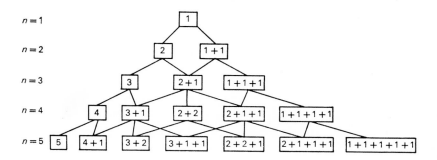

The number of partitions on the nth level is P_n; if a partition contains k *distinct* integers, it has k predecessors and $k + 1$ successors.

The fundamental property of the Young's lattice is contained in the next theorem.

THEOREM (Kreweras [5]). *Let* $\alpha = (\alpha_1, \alpha_2, \ldots, \alpha_m, 0, 0, \ldots)$ *and* $\beta = (\beta_1, \beta_2, \ldots, \beta_n, 0, 0, \ldots)$ *be sequences, as above, such that* $\beta \leqslant \alpha$ *(i.e.,* $\beta_i \leqslant \alpha_i$ *for each* i *). Define an r-chain between* α *and* β *to be a sequence* $\beta, \gamma^1, \gamma^2, \ldots, \gamma^r, \alpha$, *with*

$$\beta < \gamma^1 < \gamma^2 < \cdots < \gamma^r < \alpha.$$

The number $w_r(\beta, \alpha)$ of r chains between β and α is equal to the determinant of order $\max\{m, n\}$, whose general term (ith row and jth column) is

$$\binom{\alpha_i - \beta_j + r}{i - j + r}.$$

We assume, in the statement of this theorem, that for integers p and n

$$\binom{n}{p} = \begin{cases} \dfrac{n(n-1)(n-2)\cdots(n-p+1)}{(n-p)!} & \text{if } n > p, \\ 1 & \text{if } n = p, \\ 0 & \text{if } n < p. \end{cases}$$

For the proof, which uses induction on $\max\{m, n\}$, we refer the reader to Kreweras [5].

In particular, by varying the values of r in the determinant, this theorem gives

(1) the number of partitions γ with $\beta < \gamma < \alpha$ (case $r = 1$);

(2) the number of paths in the Young's lattice between β and α (case $r = \sum_i \alpha_i - \sum_j \beta_j - 1$).

COROLLARY (Kreweras [5]). *The number of standard tableaus of order n, associated with a partition $\alpha = (\alpha_1, \alpha_2, \ldots)$ of n, which contain a standard tableau of order p ($p < n$), associated with a partition $\beta = (\beta_1, \beta_2, \ldots)$ of p, is equal to the number of paths between β and α in the Young's lattice (which number is determined by the theorem).*

PROOF: Let us calculate the number of standard tableaus T of order n having a given diagram, and in which the integers $1, 2, \ldots, p$ ($p < n$) fill the boxes of a fixed standard tableau S of order p. Starting with S, T can be obtained by successively adding the integers $p + 1$, $p + 2$, $\ldots, n$. This procedure determines a sequence of partitions; for example,

$$S = \begin{array}{l} 1\,2 \\ 3 \end{array} \rightarrow \begin{array}{l} 1\,2\,4 \\ 3 \end{array} \rightarrow \begin{array}{l} 1\,2\,4 \\ 3\,5 \end{array} \rightarrow T = \begin{array}{l} 1\,2\,4 \\ 3\,5 \end{array}.$$

6

In the Young's lattice this determines the path:

$$\beta = (2, 1, 0, 0, \ldots) < (3, 1, \ldots) < (3, 2, \ldots) < \alpha = (3, 2, 1, \ldots).$$

Hence, there is a bijective correspondence between the standard tableaus T, satisfying the required conditions, and the paths between β and α. **Q.E.D.**

REFERENCES

1. E. T. BELL, "Algebraic Arithmetic." Amer. Math. Soc., New York, 1927.
2. P. BERTIER, Partages, parties, partitions. Decomptes et representations, *Metra* **6**, 103–129. (1967).
3. J. S. FRAME, G. DE B. ROBINSON, and R. M. THRALL, The Hook graphs of the symmetric groups, *Canad. J. Math.* **6**, 316–323 (1954).
4. G. TH. GUILBAUD, Un probleme leibnizien; les partages en nombres entiers. *Math. et Sci. Humaines* **17**, 13–16 (1966).
5. G. KREWERAS, Sur une classe de problemes de denombrement liés au treillis des partitions des entiers, Cahiers du BURO 6, ISUP, 9–103 (1965).
6. P. A. MACMAHON, "Combinatory Analysis." Cambridge Univ. Press, London, 1915 (1916), reprinted by Chelsea, New York, 1960.
7. F. D. MURNAGHAM, "The Theory of Group Representation." Williams and Wilkins, Baltimore, 1928.
8. G. DE B. ROBINSON, On the representation of the symmetric group, *Amer. J. Math.* **60**, 745–760 (1938).
9. D. E. RUTHERFORD, "Substitutional Analysis." Oliver and Boyd, Edinburgh, 1948.
10. C. SCHENSTED, Longest increasing and decreasing subsequences, *Canad. J. Math.* **13**, 179–191 (1961).
11. M. P. SCHÜTZENBERGER. Quelques remarques sur une construction de Schensted, *Math. Scand.* **12**, 117–128 (1963).
12. E. M. WRIGHT, Partition of multipartite number into k parts, *J. Reine Angew. Math.* **216**, 101–112 (1964).
13. A. YOUNG, On quantitative substitutional analysis, *Proc. London Math. Soc.* **34**, 361–397 (1902).
14. A. YOUNG, On qualitative substitutional analysis. *Proc. London Math. Soc.* **(2)28**, 255–292 (1927).

CHAPTER 3

INVERSION FORMULAS AND THEIR APPLICATIONS

1. DIFFERENTIAL OPERATOR ASSOCIATED WITH A FAMILY OF POLYNOMIALS

Consider a family

$$P_0(x), P_1(x), P_2(x), \ldots;$$

P_i of degree i

of polynomials in a real variable x, where $P_0(x) = 1$ and $P_n(0) = 0, n \geq 1$. Such a family is called a *normal family* of polynomials.

A *differential operator* D, *associated with the polynomials* $P_n(x)$, maps each polynomial $\varphi(x)$ onto a polynomial denoted by $D\varphi(x)$, and in addition satisfies

(1) $\quad DP_n(x) = \begin{cases} nP_{n-1}(x) & \text{if } n \neq 0, \\ 0 & \text{if } n = 0. \end{cases}$

(2) $\quad D[\lambda\varphi(x)] = \lambda \, D\varphi(x), \qquad \lambda = \text{constant}.$

(3) $\quad D[\varphi(x) + \varphi'(x)] = D\varphi(x) + D\varphi'(x).$

PROPERTY 1. *For each normal family* P_n *of polynomials, there exists one and only one differential operator.*

PROOF: We show that every polynomial $\varphi_n(x)$ of degree n can be expressed uniquely in the form

$$\varphi_n(x) = \alpha_n P_n(x) + \alpha_{n-1}P_{n-1}(x) + \cdots + \alpha_0 P_0(x),$$

where $\alpha_n, \alpha_{n-1}, \ldots, \alpha_0$ are constants. To prove this take for α_n the coefficient of x^n in $\varphi_n(x)$ divided by the coefficient of x^n in $P_n(x)$; $\varphi_{n-1}(x) = \varphi_n(x) - \alpha_n P_n(x)$ is then, at most, of degree $n - 1$, and for α_{n-1}

take the coefficient of x^{n-1} in $\varphi_{n-1}(x)$, divided by the coefficient of x^{n-1} in $P_{n-1}(x)$. Next consider

$$\varphi_{n-2}(x) = \varphi_{n-1}(x) - \alpha_{n-1} P_{n-1}(x),$$

which determines α_{n-2}, etc.

From (1), (2), and (3),

$$D\varphi_n(x) = n\alpha_n P_{n-1}(x) + (n-1)\alpha_{n-1} P_{n-2}(x) + \cdots + 0.$$

This proves the uniqueness of the differential operator associated with the polynomials $P_n(x)$.

Moreover, the operator D, defined by this equality, satisfies (1), (2), and (3).

THEOREM (Taylor). *Let $P_n(x)$ be a normal family of polynomials. If φ is a polynomial in x of degree n, then*

$$\varphi(x) = \varphi(0) + \frac{D\varphi(0)}{1!} P_1(x) + \frac{D^2\varphi(0)}{2!} P_2(x) + \cdots + \frac{D^n\varphi(0)}{n!} P_n(x).$$

Moreover, this series expansion of $\varphi(x)$ is unique.

PROOF: We have

$$\varphi(x) = \alpha_0 P_0(x) + \alpha_1 P_1(x) + \cdots + \alpha_n P_n(x).$$

Putting $x = 0$, this becomes

$$\varphi(0) = \alpha_0.$$

Differentiating,

$$D\varphi(x) = 0 + \alpha_1 P_0(x) + 2\alpha_2 P_1(x) + \cdots + n\alpha_n P_{n-1}(x).$$

Putting $x = 0$, we obtain

$$D\varphi(0) = \alpha_1.$$

Differentiating again,

$$D^2\varphi(x) = 2\alpha_2 P_0(x) + 2 \cdot 3\,\alpha_3 P_4(x) + \cdots + n(n-1)\alpha_n P_{n-2}(x).$$

Hence,

$$D^2\varphi(0) = 2\alpha_2.$$

More generally, let

$$1 \times 2 \times 3 \times \cdots \times k = k!;$$

then

$$D^k \varphi(0) = k! \, \alpha_k.$$

This completes the proof.

Notice that the differential operator, associated with the normal family $P_n(x) = x^n$ of polynomials, is the ordinary differential d/dx, and the formula in the above theorem becomes the standard Taylor–Maclaurin expansion.

BINOMIAL FORMULA. Let y be a constant, and consider the polynomial

$$\varphi(x) = (x + y)^n.$$

By substituting the polynomials $P_n(x) = x^n$ ($P_n(0) = 0$, $n \geq 1$) in Taylor's formula, $\varphi(x)$ may be expanded as a Taylor series.

The differential operator, associated with the polynomials P_n, is the ordinary differential

$$D\varphi(x) = d\varphi/dx.$$

Then

$$D\varphi(x) = n(x + y)^{n-1},$$
$$D^2\varphi(x) = n(n - 1)(x + y)^{n-2},$$
$$D^k\varphi(x) = n(n - 1) \cdots (n - k + 1)(x + y)^{n-k}.$$

Taylor's formula becomes

$$(x + y)^n = y^n + \binom{n}{1} x y^{n-1} + \binom{n}{2} x^2 y^{n-2} + \cdots + \binom{n}{n} x^n.$$

This is the "binomial formula" which may be written

$$\boxed{(x + y)^n = \sum_{k=0}^{n} \binom{n}{k} x^k y^{n-k}}$$

Δ BINOMIAL FORMULA. Let us consider Taylor's formula relative to the polynomials

$$P_n(x) = [x]_n = x(x-1)(x-2)\cdots(x-n+1) \quad (P_n(0) = 0,\ n \geqslant 1).$$

The operator Δ, defined by

$$\Delta\varphi(x) = \varphi(x+1) - \varphi(x),$$

is the differential operator associated with the polynomials P_n, since

$$\Delta[x]_n = [x+1]_n - [x]_n$$
$$= (x+1)[x]_{n-1} - (x-n+1)[x]_{n-1} = n[x]_{n-1}.$$

Using Taylor's formula, relative to the polynomials $[x]_n$, to expand the polynomial

$$\varphi(x) = [x+y]_n,$$

and observing that

$$\Delta^k\varphi(x) = n(n-1)\cdots(n-k+1)[x+y]_{n-k},$$

we have

$$\boxed{[x+y]_n = \sum_{k=0}^{n} \binom{n}{k}[x]_k[y]_{n-k}} \qquad \text{(Vandermonde's formula)}$$

∇ BINOMIAL FORMULA. Let us consider Taylor's formula relative to the polynomials

$$P_n(x) = [x]^n = x(x+1)(x+2)\cdots(x+n-1) \quad (P_n(0) = 0,\ n \geqslant 1).$$

The operator ∇, defined by

$$\nabla\varphi(x) = \varphi(x) - \varphi(x-1),$$

is the differential operator associated with the polynomials P_n, since

$$\nabla[x]^n = [x]^n - [x-1]^n$$
$$= (x+n-1)[x]^{n-1} - (x-1)[x]^{n-1} = n[x]^{n-1}.$$

Expanding the polynomial

$$\varphi(x) = [x+y]^n,$$

and observing that

$$\nabla^k \varphi(x) = n(n-1) \cdots (n-k+1)[x+y]^{n-k},$$

we obtain

$$\boxed{[x+y]^n = \sum_{k=0}^{n} \binom{n}{k} [x]^k [y]^{n-k}}$$

(Nörlund's formula)

FIRST INVERSION THEOREM. *Let $\varphi_n(x)$ and $\psi_n(x)$ be families of polynomials of degree n satisfying*

$$\varphi_n(x) = \sum_{k=0}^{n} \alpha_n^k \psi_k(x) \qquad (n = 0, 1, 2, \ldots, n_0),$$

$$\psi_n(x) = \sum_{k=0}^{n} \beta_n^k \varphi_k(x) \qquad (n = 0, 1, 2, \ldots, n_0).$$

If $a_0, a_1, a_2, \ldots, a_{n_0}, b_0, b_1, \ldots, b_{n_0}$ are numbers satisfying

$$a_n = \sum_{k=0}^{n} \alpha_n^k b_k \qquad (n = 0, 1, 2, \ldots, n_0),$$

then

$$b_n = \sum_{k=0}^{n} \beta_n^k a_k \qquad (n = 0, 1, 2, \ldots, n_0).$$

PROOF: Clearly,

$$\varphi_n(x) = \sum_{k=0}^{n} \alpha_n^k \sum_{m=0}^{n} \beta_k^m \varphi_m(x) = \sum_{m=0}^{n} \left(\sum_{k=0}^{n} \alpha_n^k \beta_k^m \right) \varphi_m(x).$$

Writing $\alpha_n^m = 0$, if $m > n$; $\beta_n^m = 0$, if $m > n$; $\delta_n^m = 0$, if $m \neq n$, $= 1$ if $m = n$; and comparing the coefficients of $\varphi_m(x)$ in the above equation,

$$\delta_n^m = \sum_{k=0}^{n_0} \alpha_n^k \beta_k^m.$$

In other words, the matrices $((\alpha_j^i))$ and $((\beta_j^i))$ are inverses of one another; therefore, the vector equation

$$\mathbf{a} = ((\alpha_j^i)) \mathbf{b}$$

is equivalent to

$$\mathbf{b} = ((\beta_j{}^i))\mathbf{a}.$$

REMARK: Let $P_n(x)$ and $Q_n(x)$ be families of polynomials that vanish at $x = 0$, $n \neq 0$, and let Δ and D be, respectively, their associated differential operators. What we have just shown is that the matrices

$$\left(\left(\begin{matrix} Q_0(0) & 0 & 0 & \cdots \\ Q_1(0) & \dfrac{\Delta Q_1(0)}{1!} & 0 & \cdots \\ Q_2(0) & \dfrac{\Delta Q_2(0)}{1!} & \dfrac{\Delta^2 Q_2(0)}{2!} & \cdots \\ \vdots & \vdots & \vdots & \end{matrix} \right) \right)$$

and

$$\left(\left(\begin{matrix} P_0(0) & 0 & 0 & \cdots \\ P_1(0) & \dfrac{D P_1(0)}{1!} & 0 & \cdots \\ P_2(0) & \dfrac{D P_2(0)}{1!} & \dfrac{D^2 P_2(0)}{2!} & \cdots \\ \vdots & \vdots & \vdots & \end{matrix} \right) \right)$$

are inverses of one another.

INVERSE BINOMIAL FORMULAS. Putting $x = y + 1$, and using the binomial formula,

$$x^n = (y + 1)^n = \sum_{k=0}^{n} \binom{n}{k} (x - 1)^k.$$

Again, using the binomial formula,

$$(x - 1)^n = \sum_{k=0}^{n} (-1)^{n-k} \binom{n}{k} x^k.$$

By the first inversion theorem, if the numbers $a_0, a_1, a_2, \ldots, b_0, b_1, b_2, \ldots$ satisfy

$\phi_k(x) = x^k$

$\psi_k(x) = (x-1)^k$

$$a_n = \sum_{k=0}^{n} \binom{n}{k} b_k,$$

then

$$b_n = \sum_{k=0}^{n} \binom{n}{k}(-1)^{n-k} a_k.$$

In particular (Formula 1 of Chapter 1, Section 10)

$$n^p = \sum_{k=0}^{n} \binom{n}{k}(k! S_p^{\,k}) \qquad \text{(assuming } S_p^{\,0} = 0\text{).}$$

Therefore,

$$(n! S_p^{\,n}) = \sum_{k=0}^{n}(-1)^{n-k}\binom{n}{k} k^p$$

or

$$\boxed{\; S_n^{\,m} = \frac{1}{m!}\sum_{k=0}^{m}(-1)^{m-k}\binom{m}{k} k^n \;} \qquad \text{(Stirling's formula)}$$

STIRLING'S INVERSE FORMULAS. We recall (Chapter 1, Section 5) that the Stirling numbers $s_n^{\,k}$ of the first kind are defined by

$$[x]_n = \sum_{k=1}^{n} s_n^{\,k} x^k,$$

and (Chapter 1, Section 10)

$$x^n = \sum_{k=1}^{n} S_n^{\,k}[x]_k.$$

By the first inversion theorem, if numbers $a_1, a_2, \ldots, b_1, b_2, \ldots$ satisfy

$$a_n = \sum_{k=1}^{n} s_n^{\,k} b_k,$$

then

$$b_n = \sum_{k=1}^{n} S_n^{\,k} a_k.$$

LAH'S INVERSE FORMULAS. Since $[-x]_n$ is a polynomial of degree n, it has a series expansion

$$[-x]_n = L_n{}^1[x]_1 + \cdots + L_n{}^n[x]_n.$$

The $L_n{}^k$'s are called *Lah numbers*; changing x into $-x$,

$$[x]_n = L_n{}^1[-x]_1 + \cdots + L_n{}^n[-x]_n.$$

Then, applying the first inversion theorem,

$$a_n = \sum_{k=1}^{n} L_n{}^k b_k \Leftrightarrow b_n = \sum_{k=1}^{n} L_n{}^k a_k.$$

2. THE MÖBIUS FUNCTION

We will now generalize the first inversion theorem. The following description of the Möbius function is due essentially to Rota [15], who is convinced that the Möbius function is a fundamental unifying principle of enumeration.

Let X be a set on which is defined an order relation $\leqslant$; then, by definition

$$x \leqslant x,$$
$$x \leqslant y, y \leqslant x \Rightarrow x = y,$$
$$x \leqslant y, y \leqslant z \Rightarrow x \leqslant z.$$

X is assumed to have a unique minimal element, denoted by 0; i.e.,

$$0 \leqslant x \quad (x \in X)$$

(if X does not contain such an element, then it can always be adjoined to X).

Finally, suppose that for all $x, y \in X$, the *segment*

$$[x, y] = \{u/u \in X, u \geqslant x, u \leqslant y\}$$

is finite. Such a set X, together with its order relation $\leqslant$, is called a *locally finite ordered set*.

For example, X might be the set of nonnegative integers with the order relation

$$x \leqslant y \Leftrightarrow x \text{ is less than or equal to } y,$$

or X might be the set of integers $\geqslant 1$ with the order relation

$$x \leqslant y \Leftrightarrow x \text{ is a divisor of } y.$$

Later, we shall come across many other examples.

As usual, write $x < y$, if $x \leqslant y$ and $x \neq y$; write $x \geqslant y$, if $y \leqslant x$. Let X be a locally finite ordered set. Let A be the set of all real-valued functions $f(x, y)$ of two variables, defined for x and y ranging over X, such that

$$f(x, y) \neq 0 \qquad \text{if} \quad x = y,$$
$$f(x, y) = 0 \qquad \text{if} \quad x \not\leqslant y.$$

A product $*$ is defined on A by

$$f * g(x, y) = \sum_{x \leqslant u \leqslant y} f(x, u)g(u, y).$$

The summation is over all u in the finite segment $[x, y]$.

The set A, together with the product $*$, is called the *group of arithmetic functions*. We now prove that it is, indeed, a "group."

PROPOSITION 1. *The product $*$ is associative, i.e.,*

$$(f * g) * h = f * (g * h).$$

PROOF: By definition

$$[(f * g) * h](x, z) = \sum_{\substack{y \\ x \leqslant y \leqslant z}} h(y, z) \sum_{\substack{u \\ x \leqslant u \leqslant y}} f(x, u)g(u, y)$$

$$= \sum_{\substack{u, y \\ x \leqslant u \leqslant y \leqslant z}} f(x, u)g(u, y)h(y, z)$$

$$= [f * (g * h)](x, z).$$

PROPOSITION 2. *The Kronecker function*

$$\delta(x, y) = \begin{cases} 1 & \text{if} \quad x = y, \\ 0 & \text{if} \quad x \neq y, \end{cases}$$

is the identity element for $*$.

PROOF: By definition

$$[f * \delta](x, y) = \sum_{x \leqslant u \leqslant y} f(x, u)\, \delta(u, y) = f(x, y).$$

PROPOSITION 3. *For each $f \in A$, there exists a left-inverse $f^{-1} \in A$ (that is, $f^{-1}f = \delta$); for a given x, $f^{-1}(x, y)$ is defined by induction on y as follows:*

(1) *if $y = x$:* $f^{-1}(x, y) = \dfrac{1}{f(x, x)};$

(2) *if $y > x$:* $f^{-1}(x, y) = \dfrac{-1}{f(y, y)} \sum_{x \leqslant u < y} f^{-1}(x, u)f(u, y).$

PROOF: From (1),

$$f^{-1} * f(x, x) = f^{-1}(x, x)f(x, x) = 1.$$

If $x < y$, from (2),

$$f^{-1} * f(x, y) = \sum_{x \leqslant u < y} f^{-1}(x, u)f(u, y) + f^{-1}(x, y)f(y, y) = 0.$$

Therefore, $f^{-1} * f = \delta$.

Now if each element α of a monoid has a left inverse α^{-1}, then α^{-1} is also the right inverse of α, since

$$\alpha\alpha^{-1} = f = \delta f = f^{-1}ff = f^{-1}\alpha\alpha^{-1}\alpha\alpha^{-1} = f^{-1}\alpha\,\delta\alpha^{-1} = f^{-1}f = \delta.$$

Therefore, Propositions 1, 2, and 3 prove that A is a group.

Furthermore, A is a ring (see Smith [18]; Carlitz [1]) possessing many remarkable properties.

The most important of these properties that concern us here is the following:

$$f = g * \alpha \Rightarrow g = f * \alpha^{-1}.$$

That is, to every function $\alpha(x, y)$ satisfying $\alpha(x, x) \neq 0$, $\alpha(x, y) = 0$ if $x \nleqslant y$; there corresponds a function $\beta(x, y)$ with the same properties, and such that

$$f(0, x) = \sum_{0 \leqslant u \leqslant x} \alpha(u, x)g(0, u)$$

implies

$$g(0, x) = \sum_{0 \leqslant u \leqslant x} \beta(u, x) f(0, u).$$

X being a locally finite set, the function

$$\xi(x, y) = \begin{cases} 1 & \text{if } x \leqslant y, \\ 0 & \text{otherwise}, \end{cases}$$

is called the *Riemann function*.

The *Möbius function* μ is defined inductively (for all $y \geqslant x$) by

$$\mu(x, x) = 1,$$
$$\mu(x, y) = - \sum_{x \leqslant t < y} \mu(x, t).$$

The functions ξ and μ are inverses of each other and, therefore, the following is a consequence of the first inversion theorem.

THEOREM (Möbius inversion [8]). *Let X be a locally finite ordered set and let $f(x)$, $g(x)$ be functions defined on X, with*

$$f(x) = \sum_{0 \leqslant u \leqslant x} g(u) \qquad (x \in X).$$

Then

$$g(x) = \sum_{0 \leqslant u \leqslant x} \mu(u, x) f(u) \qquad (x \in X).$$

EXAMPLE 1. Let X be the set of positive integers with the order relation $\leqslant$ defined by $k \leqslant n$ if "k is less than or equal to n."

Let us invert

$$f(n) = \sum_{k=1}^{n} g(k).$$

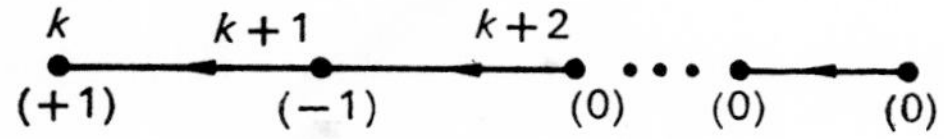

FIG. 1

For all $n \geqslant k$, the Möbius function is defined by

$$\mu(k, n) = \begin{cases} 1 & \text{if } n = k \\ -1 & \text{if } n = k + 1 \\ 0 & \text{if } n = k + 2, k + 3, \ldots . \end{cases}$$

Therefore,

$$g(n) = f(n) - f(n - 1).$$

EXAMPLE 2. Let X be the set of positive integers, with the order relation $\leqslant$ defined by

$$y \leqslant x \quad \text{if} \quad y \text{ divides } x \quad (y/x).$$

Let us find the inverse of

$$f(n) = \sum_{d/n} g(d).$$

The graph of Fig. 2 illustrates the method of finding the Möbius function $\mu(d, n)$.

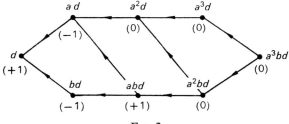

FIG. 2

It follows immediately that

$$\mu(d, n) = \begin{cases} 1 & \text{if } n = d, \\ (-1)^k & \text{if } n = p_1 p_2 \cdots p_k d, \\ 0 & \text{otherwise,} \end{cases}$$

where the p_i's are distinct prime numbers $\neq 1$. Therefore,

$$g(n) = \sum_{d/n} \mu(d, n) f(d).$$

This function $\mu(d, n)$—usually written as $\mu(d/n)$—is the classical function first derived by Möbius in 1832 in order to study the distribution of the prime numbers.

EXAMPLE 3. Let A be a finite set. Let us find the inverse of

$$f(A) = \sum_{S \subset A} g(S).$$

In this example, X is the lattice of subsets of A, with the order relation $\subset$ (inclusion).

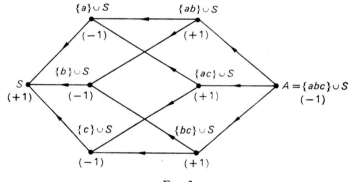

FIG. 3

Clearly,

$$\mu(S, A) = (-1)^{|A| - |S|},$$

whence,

$$g(A) = \sum_{S \subset A} (-1)^{|A| - |S|} f(S).$$

EXAMPLE 4. Let A be a set, and let

$$\mathscr{A} = (A_1, A_2, \ldots, A_k)$$

be a partition of A; then, by definition,

$$A_i \neq \varnothing,$$

$$i \neq j \Rightarrow A_i \cap A_j = \varnothing,$$

$$\bigcup A_i = A.$$

Write $\mathscr{B} < \mathscr{A}$ ($\mathscr{B}$ is a "subpartition" of $\mathscr{A}$) if

$$\left.\begin{array}{l} B_j \cap A_i \neq \varnothing \\ B_j \in \mathscr{B} \\ A_i \in \mathscr{A} \end{array}\right\} \Rightarrow B_j \subset A_i.$$

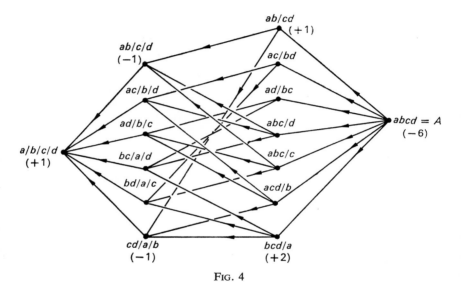

FIG. 4

If $\mathscr{A}$ is a partition of A into p classes $A_1, A_2, \ldots, A_p$, and if $\mathscr{B}$ is a subpartition of $\mathscr{A}$, such that each A_i contains n_i classes of $\mathscr{B}$, then

$$\mu(\mathscr{B}, \mathscr{A}) = (-1)^{p + n_1 + n_2 + \cdots + n_p}(n_1 - 1)!(n_2 - 1)! \cdots (n_p - 1)!.$$

This formula was first derived by Schützenberger [16] and, with a view to different applications, has since been reformulated by Frucht [23] and Rota [15].

APPLICATION: *The circular word problem* (Moreau [11]). An *alphabet* is a set of m distinct symbols $a_1, a_2, \ldots, a_m$ called *letters*; a *word of length n* is a mapping φ of $X = \{1, 2, \ldots, n\}$ into

$$A = \{a_1, a_2, \ldots, a_m\}.$$

Two words φ and φ' are said to be *equivalent*, or the same "circular word," if

$$\varphi'(i) \equiv \varphi(i + p)(\text{mod } n) \qquad (i = 1, 2, \ldots, n).$$

Let us now try to count the number of circular words of n letters.

p is said to be a *period* of a word φ if $\varphi(i + p) \equiv \varphi(i) \,(\text{mod } n)$ for each i; the *primitive period* of φ is defined to be the smallest period $p \;(1 \leqslant p \leqslant n)$. For example, the primitive period of *abcabcabc* is $p = 3$.

Obviously, a period p must be a divisor of n. Let $M(p)$ be the number of circular words of primitive period p. Then, since to each such word there corresponds p different words, the total number of words is

$$m^n = \sum_{p/n} pM(p).$$

Inverting this formula, using the Möbius function $\mu(d, n)$ of Example 2,

$$\mu(d, n) = \begin{cases} (-1)^k & \text{if } n = p_1 p_2 \ldots p_k \, d, \\ 0 & \text{otherwise}, \end{cases}$$

where the p_i's are distinct prime numbers $\neq 1$, we obtain

$$pM(p) = \sum_{q/p} \mu(q, p)m^q.$$

Therefore, the total number of circular words of length n is*

$$\sum_{p/n} M(p) = \sum_{p/n} \frac{1}{p} \sum_{q/p} \mu(q, p)m^q.$$

The problem in information theory of the "comma-free" dictionaries is a generalization of this well-known problem. A *comma-free dictionary* is a set of words, each of n letters, satisfying the property that, for any two words of the dictionary, no integer $k \;(1 \leqslant k < n)$ exists such that the last $n - k$ letters of the first word, followed by the first k letters of the second word, form a word in the dictionary. The problem is to find a comma-free dictionary with the maximum number of words.

There are several theorems that facilitate the calculation of the Möbius function in special cases, such as:

---

* Pólya's theorem, which is proved directly in Chapter 5, can be proved by an extension of this same formula; see Rota, "Enumeration under Group Action," which appeared in the *Journal of Combinatorial Theory*, 1969.

THEOREM (P. Hall). *Let the ordered set* $(X, \leqslant)$ *be a semilattice—that is, there exists for all* $a, b \in X$ *a least upper bound* c (*called the union of* a *and* b *and denoted by* $a \vee b$) *such that*

$$c \geqslant a, b,$$

$$x \geqslant a, b \Rightarrow x \geqslant c.$$

If $x > y$, *and* x *is not the union of immediate predecessors of* y, *then*

$$\mu(y, x) = 0.$$

PROOF: Assume that x is not the union of immediate predecessors of y, that is, of elements of the set

$$\{z/z \in X, z \geqslant y, \quad \text{there exists no } t \text{ such that } z > t > y\}.$$

Let $a_1, a_2, \ldots, a_k$ be the immediate predecessors of y, which are $\leqslant x$, and let $b = a_1 \vee a_2 \vee \cdots \vee a_k$, then

$y \leqslant b \leqslant x$ (since $a, a' \leqslant x$ implies $a \vee a' \leqslant x$),

$b \neq x$ (otherwise, x would be the union of predecessors of a),

$b \neq y$ (since the graph contains no circuits).

Now assume that the theorem is true for $\mu(y, z)$ with $z < x$, then it is true for $\mu(y, x)$, since

$$-\mu(y, x) = \sum_{z \in [y, x)} \mu(y, z) = \sum_{y \leqslant z < b} \mu(y, z) + \mu(y, b) + 0 = 0. \quad \textbf{Q.E.D.}$$

(This result may be verified in the above examples.)

3. SIEVE FORMULAS

Let X be a finite set on which is defined, for all $x \in X$, a function $m(x) \geqslant 0$ called the *measure* (or the *weight*) of x. If $A \subset X$, write

$$m(A) = \sum_{x \in A} m(x) \qquad \text{if} \quad A \neq \varnothing,$$

$$m(\varnothing) = 0 \qquad\qquad \text{if} \quad A = \varnothing,$$

$m(A)$ is called the *measure of the set* A.

For example, if $m(x) = 1$ for all $x \in A$, then $m(A) = |A|$ is the *cardinality* of the set A; if $m(x)$ is a probability distribution and A is the set of events having a certain property, then $m(A)$ is the probability of such an event.

FORMULA 1. *If* $A, B \subset X$ *and* $\bar{A}$ *denotes the complement* $X - A$, *then*

$$m(\bar{A}) = m(X) - m(A)$$

$$m(\overline{A \cup B}) = m(\bar{A} \cap \bar{B})$$

$$m(\overline{A \cap B}) = m(\bar{A} \cup \bar{B})$$

The proof is immediate.

FORMULA 2. *Let* A_i $(i \in \{1, 2, \ldots, q\} = Q)$ *be subsets of* X, *and let*

$$\bar{m}(K) = m\left(\bigcup_{i \in K} A_i\right) \quad \text{if} \quad K \neq \varnothing, \qquad = 0 \quad \text{if} \quad K = \varnothing,$$

$$\underline{m}(K) = m\left(\bigcap_{i \in K} A_i\right) \quad \text{if} \quad K \neq \varnothing, \qquad = 0 \quad \text{if} \quad K = \varnothing.$$

Then

$$\bar{m}(Q) = \sum_{K \subset Q} (-1)^{|K|+1} \underline{m}(K)$$

PROOF: If $q = 2$, then obviously

$$m(A_1 \cup A_2) = m(A_1) + m(A_2) - m(A_1 \cap A_2).$$

Now, proceed by induction on $q = |Q|$; assume the theorem is true for all sets of $q - 1$ subsets $A_1, A_2, \ldots, A_{q-1}$ then, for q subsets $A_1, A_2, \ldots, A_q$,

$$m(A_1 \cup A_2 \cup \cdots \cup A_{q-1} \cup A_q)$$

$$= m(A_1 \cup \cdots \cup A_{q-1}) + m(A_q) - m((A_1 \cup \cdots \cup A_{q-1}) \cap A_q)$$

$$= m(A_1 \cup \cdots \cup A_{q-1}) + m(A_q) - m\left(\bigcup_{i < q} A_i \cap A_q\right)$$

$$= \sum_{i<q} m(A_i) - \sum_{i<j<q} m(A_i \cap A_j)$$
$$+ \sum_{i<j<k<q} m(A_i \cap A_j \cap A_k) - \cdots$$
$$+ m(A_q) - \sum_{i<q} m(A_i \cap A_q)$$
$$+ \sum_{i<j<q} m(A_i \cap A_j \cap A_q) - \cdots$$
$$= \sum_{i\leqslant q} m(A_i) - \sum_{i<j\leqslant q} m(A_i \cap A_j)$$
$$+ \sum_{i<j<k\leqslant q} m(A_i \cap A_j \cap A_k) - \cdots,$$

which completes the induction.

FORMULA 3. *Let A_i $(i \in \{1, 2, \ldots, q\} = Q)$ be subsets of X. Then*

$$\underline{m}(Q) = \sum_{K \subset Q} (-1)^{|K|+1} \overline{m}(K)$$

PROOF: The Möbius function required here (see Example 3) is given by

$$\mu(I, J) = (-1)^{|J|-|I|}.$$

Using the Möbius inversion theorem, Formula 2 becomes

$$(-1)^{|Q|+1} \underline{m}(Q) = \sum_{K \subset Q} (-1)^{|K|-|Q|} \overline{m}(K).$$

Hence,

$$\underline{m}(Q) = \sum_{K \subset Q} (-1)^{|K|+1} \overline{m}(K) = \sum_{i\leqslant q} m(A_i) - \sum_{i<j\leqslant q} m(A_i \cup A_j) + \cdots.$$

Formula 3 can also be obtained from Formula 2 by taking complements.

SYLVESTER'S FORMULA. *Let $A_1, A_2, \ldots, A_q \subset X$. Let $\underline{m}(K) = \underline{m}(K)$, $K \neq \varnothing$, and $\underline{m}(\varnothing) = m(X)$; the measure of the set of elements of X that do not belong to any of the sets A_i is*

$$T_q^{\,0} = \sum_{k=0}^{q} (-1)^k \sum_{\substack{K \subset Q \\ |K|=k}} \underline{m}(K)$$

PROOF: Sylvester's formula can be deduced immediately from Formula 2, since

$$T_q^0 = m\left(\bigcap_{i \in Q} \bar{A}_i\right) = m\left(\overline{\bigcup_{i \in Q} A_i}\right) = m(X) - m\left(\bigcup_{i \in Q} A_i\right)$$

$$= \underline{\underline{m}}(\varnothing) - \sum_{\substack{K \subset Q \\ K \neq \varnothing}} (-1)^{|K|+1} \underline{\underline{m}}(K).$$

SIEVE FORMULA. *Let* $A_1, A_2, \ldots, A_q \subset X$; *the measure of the set of elements of* X *that belong to exactly* p *of the sets* A_i *is*

$$\boxed{T_q^p = \sum_{k=p}^{q} (-1)^{k-p} \binom{k}{p} \sum_{\substack{K \subset Q \\ |K| = k}} \underline{\underline{m}}(K)}$$

PROOF: Let $P \subset Q$ be an arbitrary subset of p elements. Replacing X by $\bigcap_{i \in P} A_i$, and A_j by $A_j \cap \bigcap_{i \in P} A_i$, in Sylvester's formula gives

$$m\left(\bigcap_{i \in P} A_i \cap \bigcap_{j \in Q-P} \bar{A}_j\right) = m\left(\bigcap_{i \in P} A_i\right) - \sum_{\substack{K \supset P \\ |K| = p+1}} m\left(\bigcap_{i \in K} A_i\right) + \cdots$$

$$= \sum_{K \supset P} (-1)^{|K|-|P|} \underline{\underline{m}}(K).$$

Hence,

$$T_q^p = \sum_{|P|=p} m\left(\bigcap_{i \in P} A_i \cap \bigcap_{j \in Q-P} \bar{A}_i\right)$$

$$= \sum_{|P|=p} \sum_{K \supset P} (-1)^{|K|-|P|} \underline{\underline{m}}(K)$$

$$= \sum_{\substack{K \subset Q \\ |K| \geqslant p}} \sum_{\substack{P \subset K \\ |P| = p}} (-1)^{|K|-|P|} \underline{\underline{m}}(K)$$

$$= \sum_{k=p}^{q} (-1)^{k-p} \sum_{\substack{K \subset Q \\ |K| = k}} \underline{\underline{m}}(K) \binom{k}{p}.$$

Notice that Sylvester's formula is a special case ($p = 0$) of the sieve formula.

The sieve formulas lead to a very general method of enumeration (named the sieve method after the "Sieve of Eratosthenes," which provides a construction for recursively cataloging the prime numbers). To enumerate the elements that do not belong to any of the sets $A_1, A_2, \ldots, A_q$, start with the set X and eliminate those elements belonging to A_1, then those belonging to A_2, and so on until $\bar{A}_1 \cap \bar{A}_2 \cap \cdots \cap \bar{A}_q$ is obtained.

APPLICATION 1 *"Problème des Rencontres"* (Montmort [24]). Let

$$\varphi = \begin{pmatrix} 1 & 2 & \cdots & n \\ a_1 & a_2 & \cdots & a_n \end{pmatrix}$$

be a permutation, i.e., a bijection from $\{1, 2, \ldots, n\}$ into itself. φ is said to admit a *coincidence* (or a "rencontre") *at i*, if $\varphi(i) = i$.

For example, the permutation

$$\varphi = \begin{pmatrix} 1 & 2 & 3 & 4 & 5 & 6 & 7 \\ 2 & 1 & 3 & 7 & 5 & 6 & 4 \end{pmatrix}$$

admits coincidences at 3, 5, and 6.

The problem is to obtain the number of permutations of order n which admit exactly p coincidences.

Let $P(n)$ be the number of permutations admitting no coincidences, and let A_i be the set of $(n - 1)!$ permutations that admit a coincidence at i.

From Sylvester's formula

$$P(n) = |\bar{A}_1 \cap \bar{A}_2 \cap \cdots \cap \bar{A}_n| = \underline{m}(\varnothing) - \sum_{|K|=1} \underline{m}(K) + \sum_{|K|=2} \underline{m}(K) - \cdots,$$

where, if $|K| = k$,

$$\underline{m}(K) = \left| \bigcap_{i \in K} A_i \right| = (n - k)!.$$

Therefore,

$$P(n) = n! - \binom{n}{1}(n - 1)! + \cdots + (-1)^k \binom{n}{k}(n - k)! + \cdots + (-1)^n \binom{n}{n},$$

i.e.,

$$\boxed{P(n) = n! \left[1 - \frac{1}{1!} + \frac{1}{2!} + \cdots + \frac{(-1)^k}{k!} + \cdots + (-1)^n \frac{1}{n!} \right]}$$

Obviously, the number of permutations admitting exactly p coincidences is

$$P^p(n) = \binom{n}{p} P(n - p).$$

Thus, the problem reduces to a calculation of the $P(n)$'s, which can be computed easily, using the following formulas:

$$P(1) = 0$$

$$P(n) = nP(n - 1) + (-1)^n,$$

from which we obtain

$P^p(n)$	$p = 0$	$p = 1$	$p = 2$	$p = 3$	$\cdots$
$n = 1$	0	1			
$n = 2$	1	0	1		
$n = 3$	2	3	0	1	
$n = 4$	9	8	6	0	
$n = 5$	44	45	20	10	
$\vdots$					

APPLICATION 2 "*Problème des Ménages*" (Touchard [20]). The "problème des ménages" asks for the number of ways $T(n)$ of seating n husbands (labeled $1, 2, \ldots, n$) and their respective wives (labeled $1, 2, \ldots, n$) at a circular table, in such a way that each man has a woman seated on either side of him, neither being his wife. Such a seating arrangement can be described by a bijection φ: Seat man 1; seat to his right woman $\varphi(1)$; seat to her right man 2; seat to his right woman $\varphi(2)$; etc. Let A_{2i-1} $(i \in \{1, 2, \ldots, n\})$ be the set of bijections φ with $\varphi(i) = \underline{i}$.

If $i \neq n$, let A_{2i} be the set of bijections φ with $\varphi(i) = \underline{i + 1}$ (for $i = n$, A_{2n} denotes the set of bijections φ with $\varphi(n) = \underline{1}$). From Sylvester's formula, the solution to the "problème des ménages" is

$$T(n) = \left| \bigcap_{i \in Q} \bar{A}_i \right| = \sum_{K \subset Q} (-1)^{|K|} \underline{\underline{m}}(K),$$

where $Q = \{1, 2, \ldots, 2n\}$.

If $|K| = k$

$$\underline{m}(K) = \left| \bigcap_{i \in K} A_i \right| = (n - k)! \qquad \text{if } K \text{ does not contain two consecutive integers from the sequence } (1, 2, \ldots, 2n, 1),$$

$$= 0 \qquad \text{otherwise.}$$

From Chapter 1 (Section 8, Fibonacci numbers), however, the number of sets K of cardinality k, not containing two consecutive integers from the sequence $(1, 2, \ldots, 2n, 1)$ is

$$f^*(2n, k) = \frac{2n}{2n - k} \binom{2n - k}{k}.$$

Hence,

$$T(n) = n! - \frac{2n}{2n - 1} \binom{2n - 1}{1} (n - 1)!$$

$$+ \frac{2n}{2n - 2} \binom{2n - 2}{2} (n - 2)! + \cdots + (-1)^n \frac{2n}{n} \binom{n}{n} 0!.$$

Similarly, from the sieve formula, the number of ways of seating n husbands and their wives alternately around a circular table, in such a way that exactly p husbands sit next to their own wives, is

$$T^p(n) = \sum_{k=p}^{2n} (-1)^{k-p} \binom{k}{p} \frac{2n}{2n - k} \binom{2n - k}{k} (n - k)!.$$

APPLICATION 3 *Counting prime numbers* (Euler). The problem here is to calculate the number $\varphi(n)$ of integers less than or equal to n $(n > 0)$ which are coprime to n.

Decomposing n into its prime factors suppose

$$n = p_1^{\alpha_1} p_2^{\alpha_2} \cdots p_q^{\alpha_q},$$

where $p_1, p_2, \ldots, p_q$ are distinct primes not equal to 1.

Let A_i be the set of integers $\leq n$ which are multiples of p_i; then

$$|A_i| = \frac{n}{p_i},$$

$$|A_i \cap A_j| = \frac{n}{p_i p_j}, \qquad i \neq j, \quad \text{etc.}$$

Take X to be the set of positive integers $\leq n$ and $m(A)$ equal to the cardinality of A; then, from Sylvester's formula,

$$\varphi(n) = |X| - \sum |A_i| + \sum |A_i \cap A_j| - \sum |A_i \cap A_j \cap A_k| + \cdots$$

$$= n - \sum_i \frac{n}{p_i} + \sum_{i<j} \frac{n}{p_i p_j} - \sum_{i<j<k} \frac{n}{p_i p_j p_k} + \cdots.$$

Finally,

$$\boxed{\varphi(n) = n\left(1 - \frac{1}{p_1}\right)\left(1 - \frac{1}{p_2}\right) \cdots \left(1 - \frac{1}{p_q}\right)} \qquad \text{(Euler's function)}$$

4. DISTRIBUTIONS

Let X be a set of n objects, some being indistinguishable from others; two objects x and y are said to be in the *same class* (or of the *same kind*) if they are indistinguishable; if the classes determine a partition of X of type $1^{\lambda_1} 2^{\lambda_2} \cdots n^{\lambda_n}$, then X is said to be a *collection of objects of type* $1^{\lambda_1} 2^{\lambda_2} \cdots n^{\lambda_n}$.

The objects are to be put into boxes $a_1, a_2, \ldots, a_m$ which, some of them being indistinguishable from others, form a collection of type $1^{\mu_1} 2^{\mu_2} \cdots m^{\mu_m}$.

A *distribution* is a mapping φ of X into A; two distributions are said to be *equivalent* if they are indistinguishable, i.e., if one can be obtained from the other by a permutation of objects of the same kind, or of boxes of the same kind; the classes of this equivalence relation are called *schemata*.

Let

$$R'(1^{\lambda_1} 2^{\lambda_2} \cdots n^{\lambda_n}; 1^{\mu_1} 2^{\mu_2} \cdots m^{\mu_m}).$$

be the number of schemata relative to the distributions of a collec-
tion of objects of type $1^{\lambda_1} 2^{\lambda_2} \cdots n^{\lambda_n}$ into a collection of boxes of type
$1^{\mu_1} 2^{\mu_2} \cdots m^{\mu_m}$.

Let

$$R(1^{\lambda_1} 2^{\lambda_2} \cdots n^{\lambda_n}; \ 1^{\mu_1} 2^{\mu_2} \cdots m^{\mu_m})$$

be the number of these schemata in which no box is left empty.

Then it is clear that, if $m_1 \ m_2 \cdots m_p$ denotes a partition of the integer
m into parts $m_1, m_2, \ldots, m_p$:

$$R'(1^{\lambda_1} 2^{\lambda_2} \cdots n^{\lambda_n}; \ m_1 m_2 \cdots m_p) = \sum_{\substack{0 \leqslant k_1 \leqslant m_1 \\ 0 \leqslant k_2 \leqslant m_2 \\ \vdots}} R(1^{\lambda_1} 2^{\lambda_2} \cdots n^{\lambda_n}; \ k_1 k_2 \cdots k_p).$$

Moreover, if $\mu(k_1 k_2 \cdots k_p, m_1 m_2 \cdots m_p)$ denotes the Möbius function (Chapter 3, Section 2) on the lattice of p-tuples (Chapter 1, Section
9), then

$$R(1^{\lambda_1} 2^{\lambda_2} \cdots n^{\lambda_n}; \ m_1 m_2 \cdots m_p)$$
$$= \sum_{\substack{0 \leqslant k_1 \leqslant m_1 \\ 0 \leqslant k_2 \leqslant m_2 \\ \vdots}} \mu(k_1 k_2 \cdots k_p, m_1 m_2 \cdots m_p) \times R'(1^{\lambda_1} 2^{\lambda_2} \cdots n^{\lambda_n}; \ k_1 k_2 \cdots k_p).$$

In Chapter 5, a general method for calculating R and R' will be
given; however, in a great many cases, these numbers can be calculated
directly, using very simple formulas.

PROPOSITION 1

$$R(1^n; m) = S_n^m \qquad (\textit{Stirling number}),$$
$$R'(1^n; m) = S_n^1 + S_n^2 + \cdots + S_n^m.$$

PROOF: $R(1^n; m)$ is the number of partitions of a set of n distinct
objects into m classes.

PROPOSITION 2

$$R(1^n; 1^m) = m! S_n^m$$
$$R'(1^n; 1^m) = m^n.$$

PROOF: $R(1^n; 1^m)$ is the number of surjections of X into A, and $R'(1^n; 1^m)$ is the total number of mappings of X into A.

PROPOSITION 3

$R(n; m) = P_n^m$ (*the number of partitions of n into m parts*),

$R'(n; m) = P_n^1 + P_n^2 + \cdots + P_n^m.$

PROOF: The proof is obvious.

PROPOSITION 4

$$R(n; 1^m) = \binom{n-1}{m-1},$$

$$R'(n; 1^m) = \binom{n+m-1}{n}.$$

PROOF: $R'(n; 1^m)$ is the number of solutions $u_1, u_2, \ldots, u_m$ of $u_1 + u_2 + \cdots + u_m = n$, where the u_i's are nonnegative integers. Hence (Chapter 1, Section 7)

$$R'(n; 1^m) = \frac{[n+1]^{m-1}}{(m-1)!} = \binom{n+m-1}{n}.$$

$R(n; 1^m)$ is the number of solutions $u_1, u_2, \ldots, u_m$ of $u_1 + u_2 + \cdots + u_m = n$, where the u_i's are integers greater than 0. Two solutions $(u_1, u_2, \ldots, u_m)$ and $(u_1', u_2', \ldots, u_m')$ are considered equal if and only if $u_i = u_i'$ for each i; if $s_k = u_1 + u_2 + \cdots + u_k$, each solution is completely determined by the numbers $s_1, s_2, \ldots, s_{m-1}$, satisfying

$$1 \leqslant s_1 < s_2 < \cdots < s_{m-1} \leqslant n-1,$$

and conversely.

The first formula is an obvious consequence of this remark.

THEOREM

$$R(1^{\lambda_1} 2^{\lambda_2} \cdots n^{\lambda_n}; 1^m) = \sum_{k=0}^{m} (-1)^{m-k} \binom{m}{k} \binom{k}{1}^{\lambda_1} \binom{k+1}{2}^{\lambda_2} \cdots \binom{k+n-1}{n}^{\lambda_n},$$

$$R'(1^{\lambda_1}2^{\lambda_2}\cdots n^{\lambda_n};1^m) = \binom{m}{1}^{\lambda_1}\binom{m+1}{2}^{\lambda_2}\cdots \binom{m+n-1}{n}^{\lambda_n}.$$

PROOF: Assume that some of the boxes may be left empty. First distribute the n_1 objects of the first kind, then the n_2 objects of the second kind, etc., giving as the number of possibilities

$$R'(n_1 n_2 \cdots ; 1^m) = R'(n_1; 1^m)R'(n_2; 1^m)\cdots ,$$

$$= \binom{n_1+m-1}{n_1}\binom{n_2+m-1}{n_2}\cdots .$$

This proves the second formula.

We now prove the first formula. Let R_k or $R(K)$ denote the number of distributions into a subset K ($|K| = k$) of the set A of boxes such that no box is left empty, then

$$R_m' = \sum_{K \subset A} R(K) = \sum_{k=0}^{m} \binom{m}{k} R_k .$$

Therefore, by the inverse binomial formula (Chapter 3, Section 1),

$$R_m = \sum_{k=0}^{m} (-1)^{m-k} \binom{m}{k} R_k' ,$$

whence the first formula.

APPLICATION: *The number of perfect partitions of an integer s.*
Consider a box of weights containing

t_1 weights of one kilogram,

t_2 weights of two kilograms, etc.,

where $t_1 + t_2 + \cdots = s$.

The partition $1^{t_1}2^{t_2}\cdots s^{t_s}$ is said to be *perfect*, if the box of weights will weigh any object of integral weight from 1 to s kilograms in one and only one way.

The problem is to find the number of perfect partitions of the integer s containing exactly m types of weights.

RESULT. *Let the prime decomposition of* $s + 1$ *be given by*

$$s + 1 = \prod_{i=0}^{\lambda_1} p_i{}^1 \left(\prod_{i=0}^{\lambda_2} p_i{}^2 \right)^2 \left(\prod_{i=0}^{\lambda_3} p_i{}^3 \right)^3 \cdots,$$

where for each k, $p_0{}^k = 1$, *and where for each* $i \neq 0$, *the* $p_i{}^k$ *are distinct prime numbers different from 1.*

The number of perfect partitions of s, *which consist of exactly* m *types of weights, is equal to* $R(1^{\lambda_1} 2^{\lambda_2} \cdots n^{\lambda_n}; 1^m)$.

PROOF: Suppose

$$s + 1 = p_1' p_2' \cdots p_n',$$

where the p_i's are prime numbers (not necessarily distinct) not equal to 1.

Consider a perfect partition of s consisting of exactly m types of weights. At least one of the parts of this partition must be equal to 1; if the number of 1's in this partition is $a_1 - 1$, then the succeeding part of the partition must be equal to a_1; if the number of parts equal to a_1 is $a_2 - 1$, then the succeeding part of the partition must be equal to $a_1 a_2$, and suppose there are $a_3 - 1$ of them; etc.

Therefore,

$$s = a_1 - 1 + (a_2 - 1)a_1 + (a_3 - 1)a_1 a_2 + (a_4 - 1)a_1 a_2 a_3 + \cdots$$
$$+ (a_m - 1)a_1 a_2 \cdots a_{m-1},$$

i.e.,

$$s + 1 = a_1 a_2 \cdots a_m = p_1' p_2' \cdots p_n'.$$

Such a sequence $(a_1, a_2, ..., a_m)$ completely determines a perfect partition of s consisting of exactly m types of weights; as the p''s are prime numbers not equal to 1, the number of such sequences is

$$R(1^{\lambda_1} 2^{\lambda_2} \cdots n^{\lambda_n}; 1^m). \quad \text{Q.E.D.}$$

5. COUNTING TREES

We recall some standard graph-theoretic definitions: a *graph* is defined by a set X of *vertices* and a set of *arcs*, or pairs (x, y) of vertices;

an *edge* is a set $\{x, y\}$ of two vertices joined by an arc. Let U denote the set of arcs.

A *path* is a sequence of arcs such that the terminal vertex of each arc is the initial vertex of the succeeding arc; a path that starts and finishes at the same vertex is a *circuit*. Vertices a and b belong to the same *strongly connected component* if there exists a path from a to b, and a path from b to a. A *chain* is a sequence of distinct edges, each edge having one vertex in common with the preceding edge, and the other vertex in common with the succeeding edge.

A *cycle* is a chain that begins and ends at the same vertex. Vertices a and b belong to the same *connected component* if there exists a chain from a to b.

A *partial graph* of (X, U) is a graph (X, V) with $V \subset U$; a *subgraph* of (X, U) is a graph (S, V) with $S \subset X$, and V the subset of U, consisting of all the edges that join two elements of S. We recall [22] that, if G is a graph with n vertices, m arcs, and p connected components, then the *number of independent cycles* is

$$k(G) = m - n + p.$$

For a proof of this see, for instance, Berge [22].

A *tree* is a connected graph containing no cycles; alternatively:

THEOREM 1. *Let $H = (X, U)$ be a graph of order $|X| = n \geqslant 2$; the following properties are equivalent and each characterizes a tree:*

(1)   *H is connected and contains no cycles;*

(2)   *H contains no cycles and has exactly $n - 1$ edges;*

(3)   *H is connected and has exactly $n - 1$ edges;*

(4)   *G contains no cycle and if an edge is adjoined to H one and only one cycle is created;*

(5)   *H is connected and if any edge is deleted from H, then H becomes disconnected;*

(6)   *every pair of vertices of H is connected by one and only one chain.*

PROOF:

$(1) \Rightarrow (2)$ since, if p is the number of connected components, and m the number of edges,

$$p = 1, \qquad k(H) = m - n + p = 0.$$

Therefore, $m = n - p = n - 1$.

$(2) \Rightarrow (3)$ since $k(H) = 0$, and $m = n - 1$ imply

$$p = k(H) - m + n = 1,$$

and H is connected.

$(3) \Rightarrow (4)$ since $p = 1$, and $m = n - 1$ imply

$$k(H) = m - n + p = 0.$$

Therefore, H contains no cycles; moreover, adding one edge to H increases $k(H)$ by 1, i.e., there exists in the new graph, one and only one cycle.

$(4) \Rightarrow (5)$ since, if H was not connected, there must exist two vertices a and b that are not connected, and, therefore, a new cycle could not possibly be created by adjoining the edge $\{a, b\}$; therefore, $p = 1$, $k(H) = 0$, which imply $m = n - 1$. On the other hand, deleting an edge

$$m' = n' - 2, \qquad k(H') = 0,$$

and

$$p' = k(H') - m' + n' = 2,$$

i.e., H' is disconnected.

$(5) \Rightarrow (6)$. For any two vertices a and b, there exists a chain from a to b (H is connected). If there existed a second chain from a to b, then the deletion of any edge that belonged only to the second chain would not disconnect the graph.

$(6) \Rightarrow (1)$ for if H contains a cycle, at least one pair of vertices must be connected by two distinct chains.

COROLLARY. *A graph $G = (X, U)$ possesses a partial subgraph which is a tree, if and only if, it is connected.*

PROOF: If G is not connected, no partial graph of G is connected, and, therefore, G possesses no partial trees.

If G is connected, suppose there exists an edge, whose deletion does not disconnect G. If no such edge exists then G is a tree, by (5); if such an edge exists, delete it. Now repeat this process, deleting if possible, another edge that does not disconnect G, etc.

Eventually, it will not be possible to delete any more edges without disconnecting the graph, and, hence, we shall have a tree whose set of vertices is X.

This corollary gives a simple algorithm for constructing a partial tree H of a connected graph G.

Given a graph (X, U) the *degree* of a vertex $a \in X$ is, by definition, the number $d(a)$ of edges which have a as end vertex. A vertex x, with $d(x) = 1$, is called *pendant*.

THEOREM 2 (Moon [9]). *Let* $T(n; d_1, d_2, \ldots, d_n)$ *be the number of trees with vertices* $x_1, x_2, \ldots, x_n$ *and degrees* $d(x_1) = d_1$, $d(x_2) = d_2, \ldots, d(x_n) = d_n$; *then*

$$T(n; d_1, d_2, \ldots, d_n) = \binom{n - 2}{d_1 - 1, d_2 - 1, \ldots, d_n - 1}.$$

PROOF:

(1)   Clearly, from the definitions, the sum of the degrees of any graph is twice the number of edges. Therefore, by Theorem 1, for a tree, the sum of the degrees is $2(n - 1)$. Hence, $T \neq 0$ only if

$$\sum_{i=1}^{n} (d_i - 1) = 2(n - 1) - n = n - 2.$$

We may suppose that $d_1 \geqslant d_2 \geqslant \cdots \geqslant d_n$, and otherwise relabel; since the above equality implies that $d_n = 1$, x_n is a "pendant" vertex of the tree.

(2)   In order to prove that

$$T(n; d_1, d_2, \ldots, d_n) = \sum_{\substack{i \\ d_i \geqslant 2}} T(n - 1; d_1, d_2, \ldots, d_i - 1, \ldots, d_{n-1}),$$

make a list $\mathscr{C}_i$ of the trees with vertices $x_1, x_2, \ldots, x_n$ and degrees $d(x_k) = d_k$, such that the pendant vertex x_n is joined to x_i. If $d_i \geqslant 2$,

$$|\mathscr{C}_i| = T(n - 1; d_1, d_2, \ldots, d_i - 1, \ldots, d_{n-1}).$$

Since the list of the required trees is the union of these lists $\mathscr{C}_i$ $(d_i \geqslant 2)$, the equality (2) follows.

(3) The theorem is trivially true for $n = 3$; therefore, assume $n \geqslant 3$, and suppose the theorem is true for $n - 1$; then

$$T(n; d_1, d_2, \ldots, d_n) = \sum_{\substack{i \\ d_i \geqslant 2}} T(n - 1; d_1, d_2, \ldots, d_i - 1, \ldots, d_{n-1})$$

$$= \sum_{\substack{i \\ d_i \geqslant 2}} \binom{n - 3}{d_1 - 1, d_2 - 1, \ldots, d_i - 2, \ldots, d_{n-1} - 1}$$

$$= \binom{n - 2}{d_1 - 1, d_2 - 1, \ldots, d_{n-1} - 1}$$

$$= \binom{n - 2}{d_1 - 1, d_2 - 1, \ldots, d_n - 1} \qquad \text{(since } d_n = 1\text{)}$$

(by Consequence 1 of Section 9, Chapter 1).

COROLLARY 1 (Menon [7]). *The numbers* $d_1, d_2, \ldots, d_n \geqslant 1$ *are degrees of a tree, if and only if*

$$\sum_{i=1}^{n} d_i = 2(n - 1).$$

PROOF: This condition is equivalent to $T(n; d_1, d_2, \ldots, d_n) \neq 0$.

COROLLARY 2 (Cayley's formula [2]). *The number of trees with vertices* $x_1, x_2, \ldots, x_n$ *is* n^{n-2}.

PROOF: By Theorem 2 and Corollary 1, the number of trees with vertices $x_1, x_2, \ldots, x_n$ is

$$\sum_{\substack{d_1, \ldots, d_n \geqslant 1 \\ d_1 + d_2 + \cdots + d_n = 2(n-1)}} \binom{n - 2}{d_1 - 1, d_2 - 1, \ldots, d_n - 1},$$

which, from Section 9, Chapter 1, equals

$$(1 + 1 + \cdots + 1)^{n-2} = n^{n-2}.$$

COROLLARY 3 (Clarke [3]). *The number of trees with vertices* $x_1, x_2, \ldots, x_n$ *and* $d(x_1) = k$ *is*

$$\binom{n-2}{k-1}(n-1)^{n-k-1}.$$

PROOF: The required number is

$$\sum_{d_2, d_3, \ldots, d_n} \binom{n-2}{k-1, d_2-1, d_3-1, \ldots, d_n-1}$$

$$= \frac{(n-2)!}{(k-1)!(n-k-1)!} \sum_{d_2, d_3, \ldots, d_n \geq 1} \binom{n-k-1}{d_2-1, d_3-1, \ldots, d_n-1}$$

$$= \binom{n-2}{k-1}(n-1)^{n-k-1}$$

(on putting the variables equal to one in the multinomial formula).

COROLLARY 4 (Moon [9]). *Let*

$$H_1 = (X_1, U_1), H_2 = (X_2, U_2), \ldots, H_p = (X_p, U_p)$$

be disjoint trees of orders $|X_i| = n_i$ *; the number of trees of order n, having the union of the X_i's as their set of vertices and containing the H_i's as subgraphs, is*

$$T(H_1, H_2, \ldots, H_p) = n_1 n_2 \cdots n_p n^{p-2}.$$

PROOF: For the time being assume that each set X_i is "contracted" to a unique vertex $\bar{x}_i$; the number of trees $\bar{H}$ with vertices $\bar{x}_1, \bar{x}_2, \ldots, \bar{x}_p$ and $d(\bar{x}_i) = d_i$ is, by Theorem 2,

$$\binom{p-2}{d_1-1, d_2-1, \ldots, n_p-1}.$$

Clearly, each of these trees $\bar{H}$ corresponds to $(n_1)^{d_1}(n_2)^{d_2} \cdots (n_p)^{d_p}$ trees H containing the H_i's as subgraphs. Therefore,

$$T(H_1, H_2, \ldots, H_p)$$

$$= \sum_{d_1, d_2, \ldots, d_p \geq 1} \binom{p-2}{d_1-1, d_2-1, \ldots, d_p-1}(n_1)^{d_1}(n_2)^{d_2} \cdots (n_p)^{d_p}$$

$$= n_1 n_2 \cdots n_p (n_1 + n_2 + \cdots + n_p)^{p-2}.$$

The corollary follows, since $n = n_1 + n_2 + \cdots + n_p$.

COROLLARY 5 (Cayley [2]). *The number of graphs with vertices* $x_1, x_2, \ldots, x_n$, *consisting of p disjoint trees, and with* $x_1, x_2, \ldots, x_p$ *belonging to p different trees, is*

$$T'(n; p) = pn^{n-p-1}.$$

PROOF: Make a list $\mathscr{C}$ of the trees with vertices $x_0, x_1, x_2, \ldots, x_n$, such that $d(x_0) = p$; by Corollary 3,

$$|\mathscr{C}| = \binom{n-1}{p-1} n^{n-p}.$$

If $P \subset \{1, 2, \ldots, n\}$ and $|P| = p$, let $\mathscr{C}_p$ be the list of trees belonging to $\mathscr{C}$ such that, for all $i \in P$, x_i is joined to x_0. Then

$$|\mathscr{C}| = \sum_P |\mathscr{C}_P| = \sum_P T'(n; p).$$

Hence,

$$\binom{n-1}{p-1} n^{n-p} = \binom{n}{p} T'(n; p).$$

Therefore,

$$T'(n; p) = \frac{(n-1)!}{(p-1)!(n-p)!} \cdot \frac{p!(n-p)!}{n!} n^{n-p} = pn^{n-p-1}. \quad \textbf{Q.E.D.}$$

Let $X = \{x_1, x_2, \ldots, x_n\}$ be a set of vertices, and let $U = \{u_1, u_2, \ldots, u_q\}$ be a set of edges joining pairs of vertices in X. Let us now try to find the number $T(X, U)$ of trees, with vertices $x_1, x_2, \ldots, x_n$, none of the edges of which belong to U. Using the theory of determinants, there exists a general formula (see Berge [22, Chapter 16]). However, for the particular cases that we wish to consider here, it does not prove very useful. Let (X, V) be a graph with n vertices, q edges and p connected components with, respectively, $n_1, n_2, \ldots, n_p$ vertices.
Write

$$v(V) = \begin{cases} 0 & \text{if the graph } (X, V) \text{ contains a cycle,} \\ n_1 n_2 \cdots n_p & \text{otherwise.} \end{cases}$$

THEOREM 3 (Temperley [19]). *The number of trees with vertices x_1,
$x_2, \ldots, x_n$ and no edge belonging to U is*

$$T(X, U) = n^{n-2} \sum_{V \subset U} v(V) \left(\frac{-1}{n} \right)^{|V|}.$$

PROOF: If $v \in U$, let A_v be the set of trees which contain v. Suppose
$V \subset U$. If (X, V) is acyclic and has p connected components, then, by
Corollary 4, the number of trees that contain all the edges of V is

$$\left| \bigcap_{v \in V} A_v \right| = v(V) n^{p-2} = v(V) n^{n-|V|-2}.$$

If (X, V) contains a cycle, then this equality is still valid, since both sides
vanish. Therefore, by Sylvester's formula (Chapter 3, Section 3),

$$T(X, U) = n^{n-2} + \sum_{\substack{V \subset U \\ V \neq \varnothing}} (-1)^{|V|} v(V) n^{n-2-|V|} = n^{n-2} \sum_{V \subset U} v(V) \left(\frac{-1}{n} \right)^{|V|}.$$

COROLLARY 1 (Weinberg [21]). *If U is a set of q pairwise disjoint
edges,*

$$T(X, U) = n^{n-2} \left(1 - \frac{2}{n} \right)^q.$$

PROOF: In this case, when $V \subset U$,

$$v(V) = 2^{|V|},$$

and

$$T(X, U) = n^{n-2} \sum_{k=0}^{q} 2^k \binom{q}{k} \left(\frac{-1}{n} \right)^k = n^{n-2} \left(1 - \frac{2}{n} \right)^q.$$

COROLLARY 2 (O'Neil [12]). *If U consists of q edges having a common end vertex x_1,*

$$T(X, U) = n^{n-2} \left(1 - \frac{1}{n} \right)^{q-1} \left(1 - \frac{q+1}{n} \right).$$

PROOF:

$$\sum_{V \subset U} v(V) \left(\frac{-1}{n} \right)^{|V|} = \sum_{k=0}^{q} (k+1) \binom{q}{k} \left(\frac{-1}{n} \right)^k$$

$$= \sum_{k=0}^{q} \binom{q}{k} \left(\frac{-1}{n}\right)^{k} + \sum_{k-1=0}^{q-1} \left(\frac{-q}{n}\right) \binom{q-1}{k-1} \left(\frac{-1}{n}\right)^{k-1}$$

$$= \left(1 - \frac{1}{n}\right)^{q} - \frac{q}{n} \left(1 - \frac{1}{n}\right)^{q-1}$$

$$= \left(1 - \frac{1}{n}\right)^{q-1} \left(1 - \frac{1}{n} - \frac{q}{n}\right).$$

COROLLARY 3 (Austin). *If U is the set of edges joining all the possible pairs of distinct vertices in $S \subset X$ ($|S| = s$) [i.e., (S, U) is a " complete graph"], then*

$$T(X, U) = n^{n-2} \left(1 - \frac{s}{n}\right)^{s-1}.$$

PROOF: Let $\mathscr{V}_{p}$ be the family of $V \subset U$, such that (S, V) is acyclic with p connected components, then

$$\sum_{V \subset U} v(V) \left(\frac{-1}{n}\right)^{|V|} = \sum_{p=1}^{s} \left(\frac{-1}{n}\right)^{s-p} \sum_{V \in \mathscr{V}_{p}} v(V).$$

Assume $P \subset S, |P| = p$ and $V \in \mathscr{V}_{p}$, and suppose there exists an acyclic graph with vertex set S, edge set equal to V, and a vertex of P in each connected component of (S, V), then we denote this graph by (S, V, P). By Corollary 5 of Theorem 2

$$|\{(S, V, P)/V \in \mathscr{V}_{p} \text{ and } (S, V, P) \text{ exists}\}| = ps^{s-p-1}.$$

Therefore,

$$\sum_{V \in \mathscr{V}_{p}} v(V) = \sum_{V \in \mathscr{V}_{p}} |\{(S, V, P)/P \subset S, |P| = p,$$

$$\text{and } (S, V, P) \text{ exists}\}|$$

$$= |\{(S, V, P)/P \subset S, |P| = p, V \in \mathscr{V}_{p},$$

$$\text{and } (S, V, P) \text{ exists}\}|$$

$$= \sum_{\substack{P \subset S \\ |P| = p}} ps^{s-p-1} = \binom{s}{p} ps^{s-p-1}.$$

Therefore,

$$\sum v(V)\left(\frac{-1}{n}\right)^{|V|} = \sum_{p=1}^{s} \left(\frac{-1}{n}\right)^{s-p}\binom{s}{p} p s^{s-p-1}$$

$$= \sum_{p-1=0}^{s-1} \left(\frac{-s}{n}\right)^{s-p}\binom{s-1}{p-1} = \left(1 - \frac{s}{n}\right)^{s-1}.$$

COROLLARY 4 (Scoins [17]; Glickman [4]). *If the graph* (X, U) *is the union of two disjoint complete graphs* (S, V) *and* (T, W) *with* $|S| = s$ *and* $|T| = t$, *then*

$$T(X, U) = s^{t-1}t^{s-1}.$$

PROOF: From Theorem 3,

$$\frac{T(X, V \cup W)}{n^{n-2}} = \frac{T(X, V)}{n^{n-2}} \cdot \frac{T(X, W)}{n^{n-2}}.$$

Therefore, by Corollary 3 of Theorem 3,

$$T(X, U) = n^{n-2}\left(1 - \frac{s}{n}\right)^{s-1}\left(1 - \frac{t}{n}\right)^{t-1}$$

$$= (s + t)^{s+t-2}\left(\frac{s+t-s}{s+t}\right)^{s-1}\left(\frac{s+t-t}{s+t}\right)^{t-1} = s^{t-1}t^{s-1}.$$

COROLLARY 5 (Moon [9, 10]). *Let* U *be a set of* $m - 1$ *edges forming an open chain on a set* $Y \subset X$ *of* m *vertices, then*

$$T(X, U) = n^{n-2}\sum_{p=1}^{m} \binom{m+p-1}{m-p}\left(\frac{-1}{n}\right)^{m-p}.$$

PROOF: If $V \subset U$ determines a graph (Y, V) with p connected components,

$$|V| = |Y| - p = m - p.$$

If $m_1, m_2, \ldots, m_p$ are the numbers of vertices of these components, respectively, $m_1 + m_2 + \cdots + m_p = m$.

Therefore, for $|V| = m - p$, there are as many graphs (Y, V) as there are integer solutions $m_1, m_2, \ldots, m_p > 0$ of this equation; hence,

$$\sum_{V \subset U} v(V) \left(\frac{-1}{n} \right)^{|V|} = \sum_{p=1}^{m} \left(\frac{-1}{n} \right)^{m-p} \sum_{\substack{|V| = m - p \\ V \subset U}} v(V)$$

$$= \sum_{p=1}^{m} \left(\frac{-1}{n} \right)^{m-p} \sum_{\substack{m_1, m_2, \ldots > 0 \\ m_1 + m_2 + \cdots + m_p = m}} m_1 m_2 \cdots m_p.$$

This last summation is equal to the coefficient of x^m in the expansion of

$$(x + 2x^2 + 3x^3 + \cdots)^p = x^p (1 - x)^{-2p}.$$

By the binomial formula this coefficient is equal to

$$(-1)^{m-p} \frac{-2p(-2p - 1)(-2p - 2) \cdots (-2p - (m - p - 1))}{(m - p)!}$$

$$= \frac{(m + p - 1)(m + p - 2) \cdots (2p + 1)2p}{(m - p)!}$$

$$= \binom{m + p - 1}{m - p}.$$

The corollary follows.

REFERENCES

1. L. CARLITZ, Rings of arithmetic functions, *Pacific J. Math.* **14**, 1165–1171 (1964).
2. A. CAYLEY, A theorem on trees, *Quart. J. Math.* **23**, 376–378 (1889).
3. L. E. CLARKE, On Cayley's formula for counting trees, *J. London Math. Soc.* **33**, 471–475 (1958).
4. S. GLICKMAN, On the representation and enumeration of trees, *Proc. Cambridge Philos. Soc.* **59**, 509–517 (1963).
5. I. KAPLANSKY and J. RIORDAN, the problème des menages, *Scripta Math.* **12**, 113–124 (1946).
6. H. MEIER-WUNDERLI, Note on a basis of P. Hall for higher commutation in tree groups, *Comment. Helv. Math.* **26**, 1–5 (1952).
7. V. V. MENON, On the existence of trees with given degrees, *Sankhya* **26**, 63–68 (1964).

8. A. F. Möbius, Uber eine besondere Art von Umkehruhg der Reihen, *J. Reine Angew. Math.* **9**, 105–123 (1832).
9. J. W. Moon, Enumerating labeled trees, *in* "Graph Theory and Theoretical Physics" (F. Harary, ed.), pp. 261–271. Academic Press, New York, 1967.
10. J. W. Moon, On the second moment of the complexity of a graph, *Mathematika* **11**, 95–98 (1964).
11. C. Moreau, cited by E. Lucas, "Théorie des Nombres," pp. 396–397. Gauthiers-Villars, Paris, 1891.
12. P. V. O'Neill, The number of trees in a certain network, *Notices Amer. Math. Soc.* **10**, 569 (1963).
13. A. Rényi, Some remarks on the theory of trees, *Magy. Tudom. Akad. Mat. Kut. Intéz. Kölz.* **4**, 73–85 (1959a).
14. G.-C. Rota, Private communication.
15. G.-C. Rota, On the foundations of combinatorial theory I. Theory of Möbius functions, *Z. Wahrscheinlichkeitstheorie* **2**, 340–368 (1964).
16. M. P. Schützenberger, Contribution aux applications statistiques de la théorie de l'information, *Publ. Inst. Stat. Univ. Paris* **3**, 5–117 (1954).
17. H. I. Scoins, The number of trees with nodes of alternate parity, *Proc. Cambridge Philos. Soc.* **58**, 12–16 (1962).
18. D. A. Smith, Incidence functions as generalized arithmetic functions, *Duke Math.* **34**, 617–633 (1967).
19. H. N. V. Temperley, On the mutual cancellation of cluster integrals in Mayer's fugacity series, *Proc. Phys. Soc. (London)* **83**, 3–16 (1964).
20. J. Touchard, Sur un problème de permutations, *C. R. Acad. Sci.* **198**, 631–633 (1943).
21. L. Weinberg, Number of trees in a graph, *Proc. IRE* **46**, 1954–1955 (1958).
22. C. Berge, "Théorie des Graphes et ses Applications." Dunod, Paris, 1958.
23. W. R. Frucht, and G.-C. Rota, The Möbius function for partitions of a set. *Scientia (Valparaiso)* **122**, 111–115 (1963).
24. J. Touchard, Remarques sur les probabilités totales et sur le problème des rencontres, *Ann. Soc. Sci. Bruxelles* **A53,** 126–134 (1933).

CHAPTER 4

PERMUTATION GROUPS

1. Introduction

A *permutation of degree n* is a bijection

$$\varphi = \begin{pmatrix} 1 & 2 & \cdots & n \\ k_1 & k_2 & \cdots & k_n \end{pmatrix}$$

of the set X into itself.

As with all mappings, it can be looked at in two ways. If X is a sequence $1, 2, \ldots, n$ of objects, placed in given positions, to *effect the permutation φ* on these objects is to replace the object i by the object $k_i = \varphi(i)$. The resulting n-tuple, $k_1, k_2, \ldots, k_n$, is called the *rearrangement* of the sequence $1, 2, \ldots, n$ by the permutation φ.

THEOREM 1. *The permutations of degree n form a group S_n, called the symmetric group of degree n.*

PROOF: We begin the proof by defining what is meant by a "permutation group." Consider two permutations f and g of the set $X = \{1, 2, \ldots, n\}$, for example,

$$f = \begin{pmatrix} 1 & 2 & 3 & 4 & 5 \\ k_1 & k_2 & k_3 & k_4 & k_5 \end{pmatrix}, \qquad g = \begin{pmatrix} 1 & 2 & 3 & 4 & 5 \\ 2 & 1 & 5 & 3 & 4 \end{pmatrix}.$$

The *product $f \cdot g$*, of f and g, is the permutation defined by

$$f \cdot g(i) = f(g(i)).$$

It is easy to see that

$$f \cdot g = \begin{pmatrix} 1 & 2 & 3 & 4 & 5 \\ k_1 & k_2 & k_3 & k_4 & k_5 \end{pmatrix}\begin{pmatrix} 1 & 2 & 3 & 4 & 5 \\ 2 & 1 & 5 & 3 & 4 \end{pmatrix}$$

$$= \begin{pmatrix} 2 & 1 & 5 & 3 & 4 \\ k_2 & k_1 & k_5 & k_3 & k_4 \end{pmatrix} \begin{pmatrix} 1 & 2 & 3 & 4 & 5 \\ 2 & 1 & 5 & 3 & 4 \end{pmatrix}$$

$$= \begin{pmatrix} 1 & 2 & 3 & 4 & 5 \\ k_2 & k_1 & k_5 & k_3 & k_4 \end{pmatrix},$$

i.e.,

$$f \cdot g(1) = f(g1) = f(2) = k_2,$$

$$f \cdot g(2) = f(g2) = f(1) = k_1, \quad \text{etc.}$$

That is, to calculate $f \cdot g$, the order of the columns in the first factor f is changed, in such a way, that the top row is exactly the same as the second row in the factor g, then the first row of the second factor and the second row of the first factor are taken as respectively the first and second rows of $f \cdot g$.

Notice that, in general,

$$(f \cdot g) \neq (g \cdot f),$$

for example,

$$\begin{pmatrix} 1 & 2 & 3 \\ 2 & 1 & 3 \end{pmatrix} \begin{pmatrix} 1 & 2 & 3 \\ 3 & 1 & 2 \end{pmatrix} = \begin{pmatrix} 1 & 2 & 3 \\ 3 & 2 & 1 \end{pmatrix},$$

$$\begin{pmatrix} 1 & 2 & 3 \\ 3 & 1 & 2 \end{pmatrix} \begin{pmatrix} 1 & 2 & 3 \\ 2 & 1 & 3 \end{pmatrix} = \begin{pmatrix} 1 & 2 & 3 \\ 1 & 3 & 2 \end{pmatrix}.$$

The set S_n of permutations of degree n, together with multiplication as defined above, is a *group* if the following axioms are satisfied:

(I) *Associativity*: For all $f, g, h \in S_n$,

$$f \cdot (g \cdot h) = (f \cdot g) \cdot h.$$

In this case $(f \cdot g) \cdot h$ is denoted, unambiguously, by $f \cdot g \cdot h$.

(II) *Existence of an identity element*: There exists $e \in S_n$, such that

$$f \cdot e = e \cdot f = f$$

for all $f \in S_n$.

(III) *Existence of inverses*: If $f \in S_n$, there exists an *inverse* element $f^{-1} \in S_n$, such that

$$f \cdot f^{-1} = f^{-1} \cdot f = e.$$

Axiom (I) is obviously satisfied since

$$[f \cdot (g \cdot h)](i) = f\{g[h(i)]\} = [(f \cdot g) \cdot h](i).$$

The permutation

$$e = \begin{pmatrix} 1 & 2 & \cdots & n \\ 1 & 2 & \cdots & n \end{pmatrix},$$

which fixes each object, is an identity element since

$$f \cdot e(i) = f(i), \qquad e \cdot f(i) = e[f(i)] = f(i).$$

Therefore, Axiom (II) is satisfied.

The inverse of the permutation

$$f = \begin{pmatrix} k_1 & k_2 & \cdots & k_n \\ p_1 & p_2 & \cdots & p_n \end{pmatrix}$$

is the permutation

$$f^{-1} = \begin{pmatrix} p_1 & p_2 & \cdots & p_n \\ k_1 & k_2 & \cdots & k_n \end{pmatrix}$$

obtained by interchanging the two rows of f. This follows since

$$(f \cdot f^{-1})p_i = fk_i = p_i,$$
$$(f^{-1} \cdot f)k_i = f^{-1}p_i = k_i,$$

i.e.,

$$f \cdot f^{-1} = f^{-1} \cdot f = e.$$

Therefore, Axiom (III) is satisfied.

REMARK 1: If n is a nonnegative integer, the mapping f^n defined by

$$f^n(i) = \overbrace{f \cdot f \cdots f(i)}^{n}$$

is a permutation of X. f^n is called the *nth power of f*. We write

$$f^0 = e.$$

Similarly, the mapping f^{-n} defined by

$$f^{-n}(i) = \overbrace{f^{-1} \cdot f^{-1} \cdots f^{-1}(i)}^{n}$$

is a permutation of X, and is called the *nth inverse power of f.*

A subset G of S_n is *a subgroup of S_n* if G, together with the multiplication defined above, is itself a group.

We recall the following.

THEOREM 2. *A nonempty subset H of a finite group G is a subgroup of G (denoted by $H \subseteq G$) if, and only if*

$$h, h' \in H \Rightarrow h \cdot h' \in H.$$

PROOF: Suppose $h, h' \in H \Rightarrow h \cdot h' \in H$. If $g \in H$, consider the mapping of X into itself defined by

$$\varphi(h) = g \cdot h \qquad (h \in H).$$

φ is an injection since

$$g \cdot h = g \cdot h' \Rightarrow h = g^{-1}(g \cdot h) = g^{-1}(gh') = h'.$$

Since φ is an injection, $|\varphi(H)| = |H|$. Therefore, as H is finite, φ is a bijection, i.e.,

$$\varphi(H) = \{g \cdot h / \ h \in H\} = H.$$

Hence, there exists an element h_0 belonging to H, such that $g = g \cdot h_0$ and, therefore,

$$h_0 = g^{-1}g = e,$$

is the identity element of G.

Similarly there exists an element h_1 belonging to H such that $e = g \cdot h_1$, and, therefore,

$$h_1 = g^{-1}$$

is the inverse element of g in G.

Since H contains an identity element, and for all $h \in H$ there exists an inverse belonging to H, it is a subgroup of G.

Finally, if H is a subgroup of G then, by definition, $h, h' \in H \Rightarrow h \cdot h' \in H$.

EXAMPLE. Consider the group S_3 of permutations on $\{1, 2, 3\}$:

$$e = \begin{pmatrix} 1 & 2 & 3 \\ 1 & 2 & 3 \end{pmatrix}, \qquad a = \begin{pmatrix} 1 & 2 & 3 \\ 2 & 1 & 3 \end{pmatrix}, \qquad b = \begin{pmatrix} 1 & 2 & 3 \\ 3 & 2 & 1 \end{pmatrix},$$

$$c = \begin{pmatrix} 1 & 2 & 3 \\ 1 & 3 & 2 \end{pmatrix}, \quad d = \begin{pmatrix} 1 & 2 & 3 \\ 2 & 3 & 1 \end{pmatrix}, \quad f = \begin{pmatrix} 1 & 2 & 3 \\ 3 & 1 & 2 \end{pmatrix}.$$

Then

$$c \cdot f = \begin{pmatrix} 1 & 2 & 3 \\ 1 & 3 & 2 \end{pmatrix}\begin{pmatrix} 1 & 2 & 3 \\ 3 & 1 & 2 \end{pmatrix} = \begin{pmatrix} 1 & 2 & 3 \\ 2 & 1 & 3 \end{pmatrix} = a,$$

$$f \cdot c = \begin{pmatrix} 1 & 2 & 3 \\ 3 & 1 & 2 \end{pmatrix}\begin{pmatrix} 1 & 2 & 3 \\ 1 & 3 & 2 \end{pmatrix} = \begin{pmatrix} 1 & 2 & 3 \\ 3 & 2 & 1 \end{pmatrix} = b, \quad \text{etc.}$$

By inspection, the multiplication table for S_3 is

$x \cdot y$	$y =$					
	e	a	b	c	d	f
$x = e$	e	a	b	c	d	f
a	a	e	f	d	c	b
b	b	d	e	f	a	c
c	c	f	d	e	b	a
d	d	b	c	a	f	e
f	f	c	a	b	e	d

From this table,

$$\begin{aligned}
e &\quad \text{generates} \quad \{e\}, \\
a &\quad \text{generates} \quad \{e, a\}, \\
b &\quad \text{generates} \quad \{e, b\}, \\
c &\quad \text{generates} \quad \{e, c\}, \\
d &\quad \text{generates} \quad \{e, d, f\}, \\
f &\quad \text{generates} \quad \{e, d, f\}.
\end{aligned}$$

The subgroups of S_3 are illustrated in the form of a lattice by

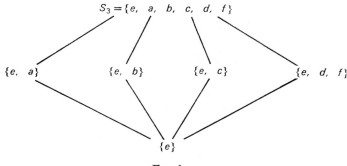

$$S_3 = \{e, \ a, \ b, \ c, \ d, \ f\}$$

$$\{e, \ a\} \qquad \{e, \ b\} \qquad \{e, \ c\} \qquad \{e, \ d, \ f\}$$

$$\{e\}$$

FIG. 1

Another important concept in group theory is that of a "normal subgroup." If G is a group, a subgroup $H \subset G$ is a *normal subgroup of G* if

$$H \cdot g = g \cdot H \qquad \text{for all} \quad g \in G,$$

where Hg denotes the set of elements of G of the form $x = hg$, $h \in H$. Equivalently, a normal subgroup H of G is a subgroup of G satisfying

$$H = gHg^{-1} \qquad \text{for all} \quad g \in G.$$

An important property of normal subgroups is: If $A \subseteq G$, $H \subseteq G$, and if H is normal in G, then,

$$A \cdot H = \bigcup_{a \in A} a \cdot H = \bigcup_{a \in A} H \cdot a = H \cdot A.$$

In other words, *the normal subgroups of G commute with the subgroups of G.*

EXAMPLE. Consider again the subgroups (Fig. 1) of $G = S_3$. Obviously, S_3 is a normal subgroup of S_3 since

$$a \cdot S_3 = \{a \cdot x / x \in S_3\} = S_3 = S_3 \cdot a.$$

Similarly, $\{e\}$ is a normal subgroup of S_3 since $a \cdot e = e \cdot a$ for all $a \in S_3$; neither $\{e, a\}$ nor $\{e, b\}$ are normal in S_3 since

$$\{e, a\} \cdot \{e, c\} = \{e, c, a, d\}; \qquad \{e, c\} \cdot \{e, a\} = \{e, a, c, f\},$$

i.e.,

$$A \cdot H \neq H \cdot A.$$

Similarly, $\{e, b\}$ is not normal in S_3. Now let us see if $\{e, d, f\}$ is normal in S_3:

$$x \cdot d \cdot x^{-1} = \begin{pmatrix} 1 & 2 & 3 \\ \alpha & \beta & \gamma \end{pmatrix} \begin{pmatrix} 1 & 2 & 3 \\ 2 & 3 & 1 \end{pmatrix} \begin{pmatrix} \alpha & \beta & \gamma \\ 1 & 2 & 3 \end{pmatrix} = \begin{pmatrix} 1 & 2 & 3 \\ \beta & \gamma & \alpha \end{pmatrix} \begin{pmatrix} \alpha & \beta & \gamma \\ 1 & 2 & 3 \end{pmatrix}$$

$$= \begin{pmatrix} \alpha & \beta & \gamma \\ \beta & \gamma & \alpha \end{pmatrix} = d \quad \text{or} \quad f;$$

$$x \cdot f \cdot x^{-1} = \begin{pmatrix} 1 & 2 & 3 \\ \alpha & \beta & \gamma \end{pmatrix} \begin{pmatrix} 1 & 2 & 3 \\ 3 & 1 & 2 \end{pmatrix} \begin{pmatrix} \alpha & \beta & \gamma \\ 1 & 2 & 3 \end{pmatrix} = \begin{pmatrix} 1 & 2 & 3 \\ \gamma & \alpha & \beta \end{pmatrix} \begin{pmatrix} \alpha & \beta & \gamma \\ 1 & 2 & 3 \end{pmatrix}$$

$$= \begin{pmatrix} \alpha & \beta & \gamma \\ \gamma & \alpha & \beta \end{pmatrix} = d \quad \text{or} \quad f.$$

Therefore, it is normal in S_3, and S_3 contains just the three normal subgroups: S_3, $\{e\}$, and $\{e, d, f\}$.

THEOREM 3. (1) *If $H \subseteq G$ is a subgroup of a group G, the sets*

$$H \cdot a = \{h \cdot a \mid h \in H\}$$

form a partition of G; the set G/H of classes of this partition is called the quotient of G, relative to H, and

$$|G/H| = |G|/|H|.$$

(2) *If H is normal in G, the quotient G/H, together with multiplication defined by*

$$K \cdot K' = \{x \cdot x' \mid x \in K, x' \in K'\},$$

is a group.

The mapping $\varphi(a) = H \cdot a$ of G into G/H is a homomorphism, i.e.,

$$\varphi(a \cdot b) = \varphi(a) \cdot \varphi(b).$$

PROOF: (1) The sets Ha form a partition of G since the relation $x \in H \cdot g$ is an equivalence relation on G:

$$x \in H \cdot x \quad \text{since} \quad x = e \cdot x,$$

$$x \in H \cdot y \implies x = h \cdot y \implies y = h^{-1} \cdot x \implies y \in H \cdot x,$$

$$\left.\begin{array}{c} x \in H \cdot y \\ y \in H \cdot z \end{array}\right\} \Rightarrow \left.\begin{array}{c} x = h \cdot y \\ y = h' \cdot z \end{array}\right\} \Rightarrow x = h \cdot h' \cdot z \Rightarrow x \in H \cdot z.$$

The class Ha has exactly $|H|$ elements since

$$h \cdot a = h' \cdot a \Rightarrow h = h' \cdot a \cdot a^{-1} = h'.$$

Therefore, the number of classes is

$$|G/H| = |G|/|H|.$$

(2) If H is normal in G, G/H is a group since

$$(Ha) \cdot (Hb) = H^2 \cdot a \cdot b = H \cdot a \cdot b,$$

$$(Ha) \cdot (He) = H^2 \cdot a = H \cdot a,$$

$$(Ha) \cdot (Ha^{-1}) = H \cdot a \cdot a^{-1} \cdot H = H^2 = H = (H \cdot e).$$

Furthermore, these formulas imply the mapping $\varphi(a) = H \cdot a$ of G into G/H is a homomorphism.

COROLLARY (Lagrange's theorem). *If $H \subseteq G$, then $|H|$ divides $|G|$.*

PROOF: From the theorem, $|G|/|H| = |G/H|$ is an integer.

For example, S_3 being of order $|S_3| = 3! = 6$, the only possible subgroups of S_3 are of order, 1, 2, 3, 6 (see the example above). However, the converse of this corollary is not true.

There exists a group G in S_4 consisting of

$$e, \quad a = \begin{pmatrix} 1 & 2 & 3 & 4 \\ 2 & 1 & 4 & 3 \end{pmatrix}, \quad b = \begin{pmatrix} 1 & 2 & 3 & 4 \\ 3 & 4 & 1 & 2 \end{pmatrix},$$

$$c = \begin{pmatrix} 1 & 2 & 3 & 4 \\ 4 & 3 & 2 & 1 \end{pmatrix}, \quad d = \begin{pmatrix} 1 & 2 & 3 & 4 \\ 2 & 3 & 1 & 4 \end{pmatrix},$$

$$f = \begin{pmatrix} 1 & 2 & 3 & 4 \\ 3 & 1 & 2 & 4 \end{pmatrix}, \quad g = \begin{pmatrix} 1 & 2 & 3 & 4 \\ 1 & 3 & 4 & 2 \end{pmatrix},$$

$$h = \begin{pmatrix} 1 & 2 & 3 & 4 \\ 1 & 4 & 2 & 3 \end{pmatrix}, \quad i = \begin{pmatrix} 1 & 2 & 3 & 4 \\ 3 & 2 & 4 & 1 \end{pmatrix},$$

$$j = \begin{pmatrix} 1 & 2 & 3 & 4 \\ 4 & 2 & 1 & 3 \end{pmatrix}, \qquad k = \begin{pmatrix} 1 & 2 & 3 & 4 \\ 2 & 4 & 3 & 1 \end{pmatrix},$$

$$l = \begin{pmatrix} 1 & 2 & 3 & 4 \\ 4 & 1 & 3 & 2 \end{pmatrix}.$$

G is of order 12, but contains no subgroup of order 6. On the other hand, it is well known that "if G is a group of order $p^m q$, p being a prime number coprime to q, then G contains a subgroup of order p^m." This is one of the Sylow theorems (it will not, however, be used below).

2. CYCLES OF A PERMUTATION

Each permutation f can be associated with a *graph* by representing the elements of X by points $1, 2, \ldots, n$, and joining the points i and $f(i)$ (and only these points) by an "arc" i.e., a line oriented from i to $f(i)$ (the direction being indicated by an arrow).

As f is a bijection, at each vertex i there exists one, and only one, incoming arc and one, and only one, outgoing arc. Now, consider the sequence $k, f(k), f^2(k), f^3(k), \ldots$, eventually an element will be repeated (since X is finite); the first such element must be k since otherwise there would exist a vertex i with at least two distinct incoming arcs.

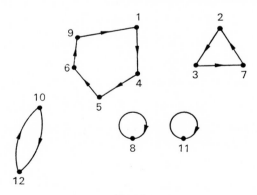

FIG. 2

EXAMPLE

$$f = \begin{pmatrix} 1 & 2 & 3 & 4 & 5 & 6 & 7 & 8 & 9 & 10 & 11 & 12 \\ 4 & 3 & 7 & 5 & 6 & 9 & 2 & 8 & 1 & 12 & 11 & 10 \end{pmatrix}.$$

Each connected component of this graph is a cycle and the different cycles partition X (see Fig. 2).

The permutation f is usually denoted by

$$f = [1 \quad 4 \quad 5 \quad 6 \quad 9] \, [2 \quad 3 \quad 7] \, [10 \quad 12] \, [8] \, [11].$$

This notation obviously describes f completely.

A permutation f, consisting of just one cycle of length greater than one, is said to be *circular*. The length of this cycle is the *length* of f.

For example,

$$f = \begin{pmatrix} 1 & 2 & 3 & 4 & 5 & 6 \\ 3 & 1 & 2 & 4 & 5 & 6 \end{pmatrix}$$

is a circular permutation on the elements 1, 2, 3. The permutation f is written

$$f = [1 \quad 3 \quad 2] \, [4] \, [5] \, [6] \qquad \text{or, more briefly,} \qquad f = [1 \quad 3 \quad 2].$$

REMARK: This notation is very convenient for computing the product of several permutations directly.

For example, consider the permutations

$$f = [1 \quad 3 \quad 4] \, [2 \quad 6],$$
$$g = [1 \quad 5 \quad 2] \, [3 \quad 6 \quad 4],$$
$$h = [1 \quad 4 \quad 5 \quad 6].$$

We shall compute the product

$$f \cdot g \cdot h = [1 \quad 3 \quad 4] \, [2 \quad 6] \, [1 \quad 5 \quad 2] \, [3 \quad 6 \quad 4] \, [1 \quad 4 \quad 5 \quad 6].$$

Consider the digit 1, and moving through the factors from right to left, the sequence of transformations that it undergoes:

$$1 \to 4 \to 3 \to 3 \to 3 \to 4.$$

Write $[1 \quad 4 \quad \cdots]$, and consider the same sequence of transformations of 4:

$$4 \to 5 \to 5 \to 2 \to 6 \to 6.$$

Write $[1 \quad 4 \quad 6 \quad \cdots]$, and consider the same sequence of transformations of 6:

$$6 \to 1 \to 1 \to 5 \to 5 \to 5.$$

Then

$$5 \to 6 \to 4 \to 4 \to 4 \to 1,$$

and the first cycle is $[1 \ 4 \ 6 \ 5]$; now take the smallest integer not included in this cycle (in this case, 2), and construct the second cycle in exactly the same way, i.e.,

$$2 \to 2 \to 2 \to 1 \to 1 \to 3,$$
$$3 \to 3 \to 6 \to 6 \to 2 \to 2.$$

Hence,

$$fgh = [1 \quad 4 \quad 6 \quad 5] [2 \quad 3].$$

Let G be a subgroup of S_n. Permutations s and t are said to be *conjugate in* G if there exists an element $g \in G$ such that $s = gtg^{-1}$. " Conjugacy " is an equivalence relation on G since

(1) reflexive: $s = ese^{-1}$; $e \in G$;
(2) symmetric: $s = gtg^{-1} \Rightarrow t = g^{-1}sg = hsh^{-1}$; $h = g^{-1} \in G$;
(3) transitive: $\left. \begin{array}{l} s = gtg^{-1} \\ t = huh^{-1} \end{array} \right\} \Rightarrow s = ghuh^{-1}g^{-1} = (gh)u(gh)^{-1}$;

$$gh \in G.$$

Some of the properties of the conjugacy classes of G will be discussed below.

THEOREM 1. *Two permutations s and t are conjugate in S_n if, and only if, they have the same number of cycles of each length.*

PROOF: (1) Suppose

$$t = [a_{11} \quad a_{12} \quad \cdots \quad a_{1i}] [a_{21} \quad a_{22} \quad \cdots \quad a_{2j}] \quad \cdots \quad [a_{m1} \quad a_{m2} \quad \cdots \quad a_{mk}].$$

Let $s = gtg^{-1}$, where g is the permutation belonging to S_n defined by $b_{pq} = g(a_{pq})$; then

$$s(b_{11}) = gtg^{-1}(b_{11}) = g \cdot t(a_{11}) = g(a_{12}) = b_{12};$$

and, therefore,

$$s = [b_{11} \ b_{12} \ \cdots \ b_{1i}] \ [b_{21} \ b_{22} \ \cdots \ b_{2j}] \ \cdots \ [b_{m1} \ b_{m2} \ \cdots \ b_{mk}].$$

Thus, two conjugate permutations have the same number of cycles of each length.

(2) Conversely, if s and t have the same number of cycles of each length (s and t are as in (1) say), then $s = gtg^{-1}$, i.e., s and t are conjugate in S_n.

EXAMPLE. S_3 contains three conjugate classes corresponding to the three partitions of n:

$$\begin{cases} 3 = 1 + 1 + 1 \\ 3 = 1 + 2 \\ 3 = 3 \end{cases}$$

The first class consists of the permutation

$$e = \begin{pmatrix} 1 & 2 & 3 \\ 1 & 2 & 3 \end{pmatrix} = [1] \, [2] \, [3].$$

The second class consists of

$$a = \begin{pmatrix} 1 & 2 & 3 \\ 2 & 1 & 3 \end{pmatrix} = [1 \ \ 2] \, [3], \qquad c = \begin{pmatrix} 1 & 2 & 3 \\ 1 & 3 & 2 \end{pmatrix} = [2 \ \ 3] \, [1],$$

$$b = \begin{pmatrix} 1 & 2 & 3 \\ 3 & 2 & 1 \end{pmatrix} = [1 \ \ 3] \, [2].$$

The third class consists of

$$d = \begin{pmatrix} 1 & 2 & 3 \\ 2 & 3 & 1 \end{pmatrix} = [1 \ \ 2 \ \ 3], \qquad f = \begin{pmatrix} 1 & 2 & 3 \\ 3 & 1 & 2 \end{pmatrix} = [1 \ \ 3 \ \ 2].$$

CAUCHY'S FORMULA. The problem we now consider is that of finding the number of permutations in a conjugacy class. More precisely, we want to find the number of permutations consisting of λ_1 cycles of length 1, λ_2 cycles of length 2, ..., λ_k cycles of length k.

In other words, we want to find the number $h(\lambda_1, \lambda_2, \ldots, \lambda_k)$ of permutations of *type* $1^{\lambda_1} \, 2^{\lambda_2} \cdots k^{\lambda_k}$.

All the permutations, with the required cycle structure, can be constructed in the following way:

$$f = \underbrace{[\times][\times] \quad \cdots \quad [\times]}_{\lambda_1} \underbrace{[\times \quad \times][\times \quad \times] \quad \cdots \quad [\times \quad \times]}_{\lambda_2}$$

$$\underbrace{\cdots\cdots \quad [\times \quad \times \quad \overbrace{\times \quad \times}^{k}]}_{\lambda_k}.$$

In this schema there are n crosses which can be replaced by the symbols $1, 2, \ldots, n$ in $n!$ ways. However, the resulting permutations are not all distinct since the λ_i cycles of length i may be permuted among themselves without changing the permutation f. Therefore, the total number of duplications of f, occurring in this way, is $\lambda_1! \, \lambda_2! \cdots \lambda_k!$.

Again, since the initial symbol of each cycle of length i can be chosen in i different ways, every cycle of length i can be written in i different ways without changing the permutation f. The number of duplications of f, occurring in this way, is $1^{\lambda_1} \, 2^{\lambda_2} \cdots k^{\lambda_k}$. Therefore, the number of distinct permutations of type $1^{\lambda_1} \, 2^{\lambda_2} \cdots k^{\lambda_k}$ is

$$\boxed{h(\lambda_1, \lambda_2, \ldots, \lambda_k) = \frac{n!}{1^{\lambda_1} \cdot \lambda_1! \, 2^{\lambda_2} \cdot \lambda_2! \cdots k^{\lambda_k} \cdot \lambda_k!}} \qquad \text{(Cauchy's formula)}$$

For example, when $n = 3$, the number of permutations of type $1 \cdot 2$ is

$$h(1, 1, 0) = \frac{3!}{1 \cdot 1! \cdot 2 \cdot 1!} = 3.$$

These permutations are (see the example above)

$$a = [1 \quad 2][3], \qquad b = [1 \quad 3][2], \qquad c = [3 \quad 2][1].$$

3. ORBITS OF A PERMUTATION GROUP

In this section, we generalize the idea of a cycle of a permutation. If $G \subseteq S_n$ is a permutation group (acting on a finite set X, $|X| = n$) and $x, y \in X$, write

$$x \equiv y \qquad (G),$$

if there exists $g \in G$ such that $y = g(x)$.

In this case, "x is equivalent to y, relative to G."

This relation $\equiv$ is an equivalence relation, since

(1) reflexive: $x \equiv x$, since $x = e(x)$;
(2) symmetric: $x \equiv y \Rightarrow y = g(x) \Rightarrow x = g^{-1}(y) \Rightarrow y \equiv x$;
(3) transitive: $\left. \begin{array}{l} x \equiv y \\ y \equiv z \end{array} \right\} \Rightarrow \left. \begin{array}{l} y = g(x) \\ z = g'(y) \end{array} \right\} \Rightarrow z = g' \cdot g(x) \Rightarrow x \equiv z.$

The equivalence classes of $\equiv$ are the *orbits* of G. If, for example, G is the subgroup $\{e, f, f^2, f^3, \ldots\}$, generated by a permutation f, the orbits of G are the cycles of f; for this reason, an orbit is a generalization of the concept of a cycle. *We shall now try to find the number of orbits of a group G.*

For all $k \in X$, let

$$G_k = \{g / g \in G, g(k) = k\},$$

i.e., G_k is the set of permutations belonging to G which fix k. By Theorem 2 (Chapter 4, Section 1), G_k is a subgroup of G, since

$$f, g \in G_k \Rightarrow f \cdot g(k) = f(k) = k \Rightarrow f \cdot g \in G_k.$$

THEOREM 1. *If O_k is the orbit of G containing k, and if G_k is the subgroup of G leaving k invariant, then*

$$|G_k| \times |O_k| = |G|.$$

PROOF: By Theorem 3 (Chapter 4, Section 1),

$$|G/G_k| = |G|/|G_k|.$$

Now we will establish a bijection of G/G_k into O_k, thus proving that $|G/G_k| = |O_k|$.

From the introduction (Theorem 3, Section 1, in particular),

$$\left. \begin{array}{l} k = g(i) = h(i) \\ g, h \in G \end{array} \right\} \Rightarrow h \cdot g^{-1}(k) = k \Leftrightarrow h \cdot g^{-1} \in G_k \Leftrightarrow h \in G_k \cdot g$$

$$\Leftrightarrow G_k \cdot g = G_k \cdot h.$$

With each $i \in O_k$, associate the permutations g_i satisfying $g_i(i) = k$ (by the definition of O_k, at least one such permutation exists), i.e., with each i, associate the class $G_k \cdot g_i \in G/G_k$.

This defines a mapping of O_k into G/G_k which is

(1) injective (from the above);

(2) surjective, since $G_k g$ is the image of $g^{-1}(k) \in O_k$, for all $G_k \cdot g \in G/G_k$.

Thus, this mapping is a bijection and

$$|O_k| = |G/G_k|.$$

The theorem follows.

EXAMPLE. Consider the subgroup $G \subseteq S_5$ generated by $a = [1 \ 2 \ 3][4 \ 5]$.

The elements of G are

$$
\begin{aligned}
a &= [1 \quad 2 \quad 3]\,[4 \quad 5] \\
a^2 &= [1 \quad 3 \quad 2]\,[4]\,[5] \\
a^3 &= [1]\,[2]\,[3]\,[4 \quad 5] \\
a^4 &= [1 \quad 2 \quad 3]\,[4]\,[5] \\
a^5 &= [1 \quad 3 \quad 2]\,[4 \quad 5] \\
a^6 &= [1]\,[2]\,[3]\,[4]\,[5] = e.
\end{aligned}
$$

The orbits are

$$O = \{1, 2, 3\} \qquad \text{and} \qquad O' = \{4, 5\}.$$

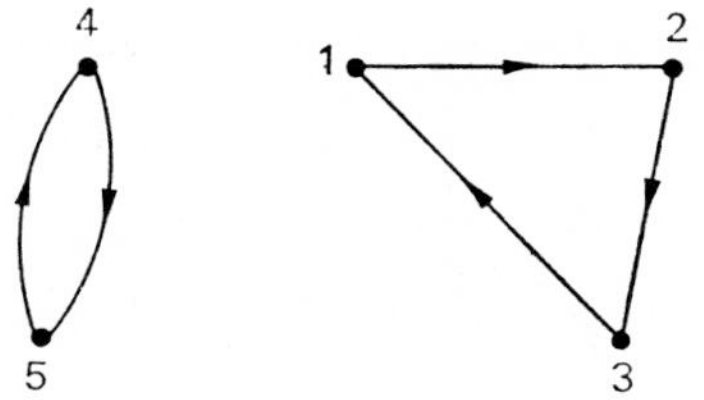

FIG. 3

Let

$$O_1 = \{1, 2, 3\},$$

then

$$G_1 = \{a^3, a^6\},$$

$$G = \{a, a^2, a^3, a^4, a^5, a^6\},$$

and, verifying Theorem 1,

$$|G_1| \times |O_1| = 3 \times 2 = 6 = |G|.$$

THEOREM 2 (Burnside's lemma). *If $\lambda_1(g)$ is the number of elements of X fixed by the permutation g (that is the number of cycles of g of length 1), then the number of orbits of a group $G \subseteq S_n$ is*

$$|\mathscr{C}_G| = \frac{1}{|G|} \sum_{g \in G} \lambda_1(g).$$

PROOF: We want to count, in two different ways, the number of ordered pairs (g, k) satisfying $g(k) = k$, $g \in G$, $k \in X$, viz.,

$$\sum_{g \in G} \lambda_1(g) = \sum_{k \in X} |G_k| = \sum_{O \in \mathscr{C}_G} \sum_{k \in O} |G_k|.$$

If j and k belong to the same orbit O, then (Theorem 1 of Section 3)

$$|G_j| = |G|/|O| = |G_k|.$$

Therefore,

$$\sum_{g \in G} \lambda_1(g) = \sum_{O \in \mathscr{C}_G} |O| \cdot |G|/|O| = |G| \times |\mathscr{C}_G|. \quad \textbf{Q.E.D.}$$

In the next chapter we will see how very important, in relation to counting problems, this theorem is.

EXAMPLE. Applying Theorem 2 to the preceding example,

$$\lambda_1(a) = 0, \qquad \lambda_1(a^2) = 2, \qquad \lambda_1(a^3) = 3,$$
$$\lambda_1(a^4) = 2, \qquad \lambda_1(a^5) = 0, \qquad \lambda_1(a^6) = 5,$$
$$|\mathscr{C}_G| = \tfrac{1}{6}(2 + 3 + 2 + 5) = 2.$$

4. PARITY OF A PERMUTATION

Consider a permutation, for example,

$$g = \begin{pmatrix} 1 & 2 & 3 & 4 & 5 \\ 4 & 5 & 1 & 2 & 3 \end{pmatrix} = 4 \quad 5 \quad 1 \quad 2 \quad 3.$$

The integer 1 is said to *introduce two inversions* in g, because two of the integers preceding 1 are greater than 1.

Similarly,

> 2   introduces two inversions: 4 and 5,
>
> 3   introduces two inversions: 4 and 5,
>
> 4   introduces no inversions,
>
> 5   introduces no inversions.

Therefore, the *total number of inversions* of g is

$$I(g) = 2 + 2 + 2 = 6.$$

The *signature $p(g)$ of g* is defined by

$$p(g) = (-1)^{I(g)}.$$

If $p(g) = +1$ the permutation is *even*; if $p(g) = -1$ it is *odd*.

A permutation, for example, $t = [i, j]$, consisting of just one cycle of length 2 (all the other cycles being of length 1), is called a *transposition*. Notice that a transposition t is equal to its own *inverse*.

If $g = 4\ 5\ 1\ 2\ 3$, and $t = [2, 3]$,

$$g \cdot t = \begin{pmatrix} 1 & 2 & 3 & 4 & 5 \\ 4 & 5 & 1 & 2 & 3 \end{pmatrix} \begin{pmatrix} 1 & 2 & 3 & 4 & 5 \\ 1 & 3 & 2 & 4 & 5 \end{pmatrix} = \begin{pmatrix} 1 & 2 & 3 & 4 & 5 \\ 4 & 1 & 5 & 2 & 3 \end{pmatrix}$$

$$= 4\ 1\ 5\ 2\ 3.$$

The sequence 4 5 1 2 3 becomes the sequence 4 1 5 2 3; the second and third terms are interchanged and the remaining terms are unaltered.

What is the minimum number of transpositions of the form $[i, i + 1]$ required to transform the sequence $k_1, k_2, \ldots, k_n$ into 1, 2, $\ldots, n$? For example,

$$4\ 5\ 1\ 2\ 3 \xrightarrow{[2\ 3]} 4\ 1\ 5\ 2\ 3 \xrightarrow{[1\ 2]} 1\ 4\ 5\ 2\ 3 \xrightarrow{[3\ 4]} 1\ 4\ 2\ 5\ 3$$

$$\xrightarrow{[2\ 3]} 1\ 2\ 4\ 5\ 3 \xrightarrow{[4\ 5]} 1\ 2\ 4\ 3\ 5 \xrightarrow{[3\ 4]} 1\ 2\ 3\ 4\ 5.$$

In this case, the minimum number of such transpositions is 6. The basic result is:

THEOREM 1. *The minimum number of transpositions, of the form*
$[i, i + 1]$, *required to transform the sequence* $g = k_1 k_2 \cdots k_n$ *into the*
sequence $1\ 2\ \cdots\ n$ *is equal to the number* $I(g)$ *of inversions of* g. *More-*
over, if after q *transpositions of the form* $[i, i + 1]$, *the sequence*
$k_1 k_2 \cdots k_n$ *is transformed into* $1\ 2\ \cdots\ n$, *then* q *and* $I(g)$ *have the*
same parity.

PROOF:

(1)   We show, first of all, that the sequence

$$k_1 \quad k_2 \quad \cdots \quad k_n$$

can be transformed into $1\ 2\ \cdots\ n$ by $I(g)$ of the required transpositions.
By interchanging two consecutive terms as many times as the number
of inversions introduced by 1, the symbol 1 can be brought to the first
position; next, by interchanging two consecutive terms as many times
as the number of inversions introduced by 2, the symbol 2 can be
brought to the second position; etc.

Now, with every interchange, I is reduced by 1 until eventually
$I = 0$, i.e., $k_1 k_2 \cdots k_n$ is transformed into $1\ 2\ \cdots\ n$ by $I(g)$ of the
required transformations.

(2)   The transposition $[i, i + 1]$ transforms the sequence $g =$
$k_1 k_2 \cdots k_n$ into the sequence $g' = k_1 k_2 \cdots k_{i-1} k_{i+1} k_i k_{i+2} \cdots k_n$,
where

$$I(g') = I(g) + 1 \quad \text{if} \quad k_i < k_{i+1},$$
$$= I(g) - 1 \quad \text{if} \quad k_i > k_{i+1}.$$

Hence, at least $I(g)$ such transpositions are required to transform g
into $1\ 2\ \cdots\ n$.

Therefore, by (1), $I(g)$ is the minimum number required. Moreover,
if q transpositions $[i, i + 1]$ transform g into $1\ 2\ \cdots\ n$, then, since I
is reduced from $I(g)$ to 0 in the process, the above equations imply
that q and $I(g)$ have the same parity.

COROLLARY. *The group* S_n *is generated by the* $n - 1$ *transpositions*
$[1\ 2][2\ 3]\ \cdots\ [n - 1\ n]$. *If a permutation* g *can be expressed as the*
product of q *of these transpositions then* q *and* $I(g)$ *have the same parity.*

PROOF: If the sequence $k_1 \, k_2 \, \cdots \, k_n$ can be transformed into $1 \ 2 \ \cdots \ n$ by transpositions $t_1, t_2, \ldots, t_q$, then the permutation

$$g = \begin{pmatrix} 1 & 2 & \cdots & n \\ k_1 & k_2 & \cdots & k_n \end{pmatrix}$$

satisfies

$$g \cdot t_1 \cdot t_2 \cdot t_3 \cdots t_q = e$$

or

$$g = (t_1 \quad t_2 \quad \cdots \quad t_q)^{-1} = t_q^{-1} \cdot t_{q-1}^{-1} \cdots t_1^{-1} = t_q \cdot t_{q-1} \cdots t_1.$$

The corollary follows immediately from the theorem.

THEOREM 2. *If $p(g) = (-1)^{I(g)}$ denotes the signature of a permutation g, then*

$$p(g \cdot g') = p(g) \times p(g')$$

(in other words, p is a homomorphism of S_n into the multiplicative group $\{1, -1\}$).

PROOF: Express g and g', in the way shown above, as a product of canonical transpositions. Then

$$g \cdot g' = (t_1 \cdot t_2 \cdots t_{I(g)}) \cdot (t_1' \cdot t_2' \cdots t_{I(g')}').$$

This is a product of $I(g) + I(g')$ canonical transpositions, and therefore, from the corollary, $I(g) + I(g')$ and $I(g \cdot g')$ have the same parity; whence

$$(-1)^{I(g \cdot g')} = (-1)^{I(g)} \cdot (-1)^{I(g')}.$$

COROLLARY 1. *If g can be expressed as the product of q (not necessarily canonical) transpositions, then q and $I(g)$ have the same parity.*

PROOF:

(1) We first of all prove that an arbitrary transposition $t = [i, j]$ is odd. In the sequence $1, 2, \ldots, i - 1, j, i + 1, \ldots, j - 1, i, j + 1, \ldots, n, i$ can be brought to the ith position by effecting, on this sequence succes-

sively, the $j - i$ transpositions $[j - 1, j]$, $[j - 2, j - 1]$, $[j - 3, j - 2]$, $\ldots$, $[i, i + 1]$. This leaves j in the $(i + 1)$th position, and j can now, in the same way, be brought to the jth position by effecting $j - (i + 1)$ canonical transpositions on the sequence. Therefore, $2(j - i) - 1$ canonical transpositions transform the above sequence into $1\ 2\ \cdots\ n$. Hence, $p(t) = -1$.

(2) Let $g = t_1 \cdot t_2 \cdots t_q$, where $t_1, t_2, \ldots, t_q$ are arbitrary transpositions.

By the theorem and (1),

$$(-1)^{I(g)} = p(g) = p(t_1) \times p(t_2) \times \cdots \times p(t_q) = (-1)^q.$$

Therefore, q and $I(g)$ have the same parity.

COROLLARY 2. *In any permutation group $G \subseteq S_n$, either all the permutations are odd, or there are as many odd permutations as even permutations.*

PROOF: Suppose G contains an odd permutation h, i.e., $p(h) = -1$; then, since the mapping $g \to hg$ of G into itself is a bijection,

$$\sum_{g \in G} p(g) = \sum_{g \in G} p(hg) = \sum_{g \in G} p(h)p(g) = -\sum_{g \in G} p(g).$$

Therefore,

$$\sum_{g \in G} p(g) = 0.$$

Hence, the number of permutations g with $p(g) = +1$ is equal to the number of permutations g with $p(g) = -1$.

COROLLARY 3. *If g is a permutation of type $1^{\lambda_1}\ 2^{\lambda_2} \cdots n^{\lambda_n}$, the parity of g is equal to the parity of $\lambda_2 + \lambda_4 + \lambda_6 + \cdots$; in other words, the parity of a permutation is the same as the parity of the number of its even cycles.*

PROOF:

(1) We show, first of all, that a cycle of length m can be expressed as a product of $m - 1$ transpositions.

Consider the circular permutation

$$g = \begin{pmatrix} k_1 & k_2 & \cdots & k_m \\ k_2 & k_3 & \cdots & k_1 \end{pmatrix}.$$

Then

$$g = [k_1 \ k_2] \cdot [k_2 \ k_3] \ \cdots \ [k_{m-1} \ k_m],$$

which is a product of $m - 1$ transpositions.

(2) By (1), if g is of type $1^{\lambda_1} 2^{\lambda_2} \cdots n^{\lambda_n}$, it can be expressed as a product of $\lambda_2 + 2\lambda_3 + 3\lambda_4 + \cdots$ transpositions; therefore, from Corollary 1, g and $\lambda_2 + \lambda_4 + \lambda_6 + \cdots$ have the same parity.

COROLLARY 4. *The set of even permutations*

$$A_n = \{g/ \ g \in S_n, \ p(g) = +1\}$$

is a normal subgroup, called the alternating group, of S_n. A_n contains $\frac{1}{2}n!$ elements.

PROOF: First of all, we prove a more general result, viz., if $\bar{g} = p(g)$ is a homomorphism of a group G into a group $\bar{G}$, then

$$K = \{g/ \ g \in G, p(g) = \bar{e}\}$$

(called the *kernel* of p) is a normal subgroup of G.

(1) K is a subgroup, since

$$g, g' \in K \Rightarrow p(g) = p(g') = \bar{e} \Rightarrow p(g \cdot g') = p(g) \cdot p(g') = \bar{e}$$
$$\Rightarrow g \cdot g' \in K.$$

(2) K is normal in G, since

$$x \in Kg \Rightarrow \begin{array}{l} x = kg \\ p(k) = \bar{e} \end{array} \Rightarrow p(x) = p(g)$$
$$\Rightarrow p^{-1}(g)p(x) = \bar{e} \Rightarrow p(g^{-1}x) = \bar{e}$$
$$\Rightarrow g^{-1}x \in K \Rightarrow x \in gK.$$

Therefore, $Kg \subseteq gK$. Similarly, $gK \subseteq Kg$, and hence $gK = Kg$. Thus, K is normal in G. Now, letting G be the permutation group S_n and $\bar{G}$ the multiplicative group $\{+1, -1\}$, the corollary follows immediately.

From Corollary 2, the number of elements in A_n is

$$|A_n| = \tfrac{1}{2}|S_n| = \tfrac{1}{2}n!.$$

THEOREM 3. *The alternating group A_n is generated by the $n-2$ circular permutations*

$$t_3 = [1\quad 2\quad 3], \quad t_4 = [1\quad 2\quad 4], \quad \ldots, \quad t_n = [1\quad 2\quad n].$$

PROOF: By the corollary to Theorem 1, Chapter 4, Section 4, since

$$[1\quad j]\cdot[1\quad i]\cdot[1\quad j] = [i\quad j],$$

S_n is generated by the transpositions $[1\quad 2], [1\quad 3], \ldots, [1\quad n]$.

Let $g \in A_n$; g can be expressed as a product of an even number of transpositions of the form $[i, i+1]$, and therefore, as a product of an even number of transpositions of the form $[1, j]$.

However,

$$[1\quad j]\cdot[1\quad i] = [1\quad i\quad j].$$

Therefore, A_n is generated by the circular permutations $[1\ i\ j]$. Finally,

$$[1\quad i\quad j] = [1\quad 2\quad j]\cdot[1\quad 2\quad j]\cdot[1\quad 2\quad i]\cdot[1\quad 2\quad j].$$

Therefore, A_n is generated by the circular permutations $[1\ 2\ i]$.

LEMMA 1. *If H is a normal subgroup of A_n $(n > 3)$, containing a circular permutation on three letters, then $H = A_n$.*

PROOF: Let $h = [1\ 2\ 3] \in H$; since H is normal in A_n, it contains the permutation

$$g[1\quad 2\quad 3]g^{-1} \quad (g \in A_n).$$

In particular, letting $g = [3\ 2\ k]$, $k > 3$, H contains

$$[3\quad 2\quad k]\cdot[1\quad 2\quad 3]\cdot[k\quad 2\quad 3] = [1\quad k\quad 2].$$

Therefore, H contains

$$[1\quad k\quad 2]\cdot[1\quad k\quad 2] = [1\quad 2\quad k].$$

Since, by Theorem 3, A_n is generated by the permutations of the form $[1\ 2\ k]$, this implies $H = A_n$.

LEMMA 2. *If H is a normal subgroup of A_n $(n > 3)$, and if $h \in H$ contains a cycle of length > 3, then $H = A_n$.*

PROOF: Let $h = a \cdot b \cdot c \cdots$ be the decomposition of h into disjoint cycles, and assume that one of these cycles has length > 3, say

$$a = [1\ \ 2\ \ \cdots\ \ m], \qquad m > 3.$$

Since $g = [1\ 2\ 3] \in A_n$ and H is normal in A_n,

$$h_1 = ghg^{-1} = (gag^{-1})bcd \cdots \in H.$$

Therefore, H also contains

$$h^{-1}h_1 = (\cdots \quad c^{-1} \quad b^{-1} \quad a^{-1})(gag^{-1})bc \quad \cdots$$
$$= a^{-1}gag^{-1}(\cdots \quad c^{-1}b^{-1})(bc \quad \cdots)$$
$$= a^{-1}gag^{-1}$$
$$= [m\ \ \cdots\ \ 3\ \ 2\ \ 1][1\ \ 2\ \ 3][1\ \ 2\ \ 3\ \ \cdots\ \ m][3\ \ 2\ \ 1]$$
$$= [1\ \ 3\ \ m].$$

Hence, by Lemma 1, $H = A_n$.

THEOREM 4 (Galois). *If $n > 4$, the only normal subgroups of A_n are A_n and $\{e\}$.*

PROOF: Let H be a normal subgroup of A_n $(n > 4)$ different from A_n and $\{e\}$. By Lemma 2, $h \in H$ can be decomposed into a product of disjoint cycles of length 3 or less; notice that, since these cycles are disjoint, the order in which they appear in the product is unimportant.

(1)   Suppose h contains two cycles of length 3; then we may suppose

$$h = [1\ \ 2\ \ 3] \cdot [4\ \ 5\ \ 6] \cdot h'.$$

As $g = [2\ 3\ 4] \in A_n$, H contains

$$h_1 = ghg^{-1} = [2 \quad 3 \quad 4]\,[1 \quad 2 \quad 3]\,[4 \quad 5 \quad 6]\,[4 \quad 3 \quad 2]h'$$
$$= [1 \quad 3 \quad 4] \cdot [2 \quad 5 \quad 6]h'.$$

H also contains

$$h^{-1}h_1 = [3 \quad 2 \quad 1]\,[6 \quad 5 \quad 4]h'^{-1}[1 \quad 3 \quad 4]\,[2 \quad 5 \quad 6]h'$$
$$= [3 \quad 2 \quad 1]\,[6 \quad 5 \quad 4]\,[1 \quad 3 \quad 4]\,[2 \quad 5 \quad 6] = [1 \quad 2 \quad 4 \quad 3 \quad 6].$$

Therefore, by Lemma 2, $H = A_n$, which is a contradiction.

(2)   If h contains exactly one cycle of length 3, then we may assume

$$h = [1 \quad 2 \quad 3]h',$$

where h' consists of disjoint cycles of length less than 3.

Therefore, H contains

$$h^2 = [1 \quad 2 \quad 3]\,[1 \quad 2 \quad 3](h')^2 = [1 \quad 3 \quad 2].$$

Hence, by Lemma 1, $H = A_n$, which is a contradiction.

(3)   If h ($h \neq e$) contains no cycle of length 3, then, since $n > 4$, we may suppose

$$h = [1 \quad 2] \cdot [3 \quad 4]h',$$

where h' consists of disjoint cycles of length less than 3.

Since $g = [2 \; 3 \; 4] \in A_n$, H contains

$$h_1 = ghg^{-1} = [2 \quad 3 \quad 4]\,[1 \quad 2] \cdot [34]\,[4 \quad 3 \quad 2]h' = [1 \quad 3]\,[2 \quad 4]h'.$$

H also contains

$$h_2 = h^{-1}h_1 = [2 \quad 1]\,[4 \quad 3]\,[1 \quad 3]\,[2 \quad 4] = [1 \quad 4]\,[2 \quad 3].$$

Since $[1 \; 4 \; 5] \in A_n$, H also contains

$$h_3 = [1 \quad 4 \quad 5]h_2[1 \quad 4 \quad 5]^{-1} = [1 \quad 4 \quad 5]\,[1 \quad 4]\,[2 \quad 3]\,[5 \quad 4 \quad 1]$$
$$= [2 \quad 3]\,[4 \quad 5].$$

Finally, H also contains

$$h_2 h_3 = [1 \quad 4]\,[2 \quad 3]\,[2 \quad 3]\,[4 \quad 5] = [1 \quad 4 \quad 5].$$

Therefore, by Lemma 1, $H = A_n$, which is a contradiction. This completes the theorem.

REMARK:

If $n = 2$, $A_n = \{e\}$.

If $n = 3$, $A_n = \{e, d, f\}$ (cf. example, Chapter 4, Section 1), and the only normal subgroups are A_n and $\{e\}$.

If $n = 4$, there exists a nontrivial normal subgroup H of A_n. We now show how to construct H.

If $h \in H$ ($h \neq e$), since h is even, by Theorem 2, Corollary 3, h contains either 0 or 2 even cycles.

By Lemma 1, h cannot be of type $1 \cdot 3$. h cannot be of type 1^4 since $h \neq e$. Therefore, H contains a permutation of type 2^2:

$$h = [1 \quad 2] [3 \quad 4] \qquad \text{(say)}.$$

H also contains

$$[1 \quad 2 \quad 3]h[1 \quad 2 \quad 3]^{-1} = [1 \quad 2 \quad 3][1 \quad 2][3 \quad 4][3 \quad 2 \quad 1]$$
$$= [1 \quad 4][2 \quad 3]$$
$$[1 \quad 3 \quad 4]h[1 \quad 3 \quad 4]^{-1} = [1 \quad 3 \quad 4][1 \quad 2][3 \quad 4][4 \quad 3 \quad 1]$$
$$= [1 \quad 4][2 \quad 3]$$
$$[2 \quad 3 \quad 4]h[2 \quad 3 \quad 4]^{-1} = [2 \quad 3 \quad 4][1 \quad 2][3 \quad 4][4 \quad 3 \quad 2]$$
$$= [1 \quad 3][2 \quad 4] \text{ etc.}$$

Finally, by inspection, H is a normal subgroup consisting of

$$e, \quad [1 \quad 2][3 \quad 4], \quad [1 \quad 3][2 \quad 4], \quad [1 \quad 4][2 \quad 3].$$

H is sometimes called the *four group* (or the *Klein group*). The Galois theorem has a very famous application (due to Evariste Galois), viz., the classical result, that every algebraic equation of degree greater than 4 is not solvable by radicals, is a consequence of this theorem.

APPLICATION: THE PERMUTOHEDRON. Consider a graph in which the vertices are the $n!$ permutations on $X = \{1, 2, \ldots, n\}$, and two vertices f and g are joined by an edge if, and only if, there exists a transposition t such that $f = tg$. This graph can be represented (Fig. 4) by a convex polyhedron, which Guilbaud and Rosenstiehl [1] have suggested calling a "permutohedron."

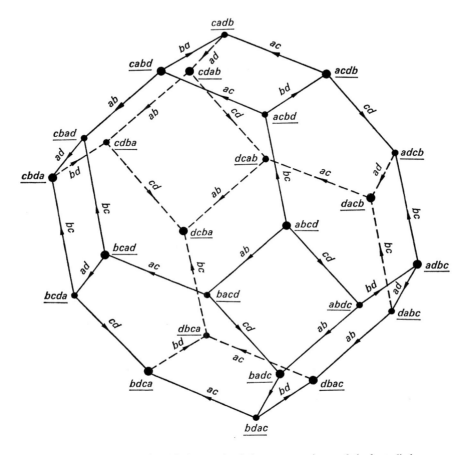

FIG. 4. Representation of the graph of the permutations of {a, b, c, d} by a convex polyhedron ("permutohedron"). (The relevant transpositions are represented by arcs.)

When two vertices $f = x_1 x_2 \cdots x_i \cdots x_j \cdots x_n$ and $g = x_1 x_2 \cdots x_j \cdots x_i \cdots x_n$ are adjacent the arc is directed from f to g if $x_i < x_j$, and from g to f if $x_j < x_i$. We shall now show, following Rosenstiehl [2], that this construction defines a lattice.

If $f = x_1 x_2 \cdots x_n$, let $E(f)$ be the set of pairs $x_i x_j$ which do not introduce an inversion $[(n(n-1)/2) - I(f)$ in number$]$; let $E^*(f)$ be the set of $I(f)$ pairs which do introduce an inversion, i.e.,

$$E(f) = \{x_i x_j \mid i < j, \, x_i < x_j\},$$
$$E^*(f) = \{x_i x_j \mid i < j, \, x_i > x_j\}.$$

Let X be a finite set. Let E be a set of pairs xy with $x, y \in X$, $x < y$. Let E^* be the set of all pairs xy, $x \in X$, $y \in X$, such that $x > y$ and $yx \notin E$. The directed graph $(X, E \cup E^*)$ is defined in the following way: the vertices are the elements of $X = \{1, 2, \ldots, n\}$, and x and y are joined by an arc directed from x to y if, and only if, $xy \in E \cup E^*$. This graph is *complete*, i.e., for any two vertices x and y ($x < y$, say), there exists an arc directed from x to y (if $xy \in E$), or an arc directed from y to x (if $yx \in E^*$). One or other of these possibilities must hold.

LEMMA. *A complete graph, containing no circuits, contains a path $a_1, a_2, \ldots, a_n$ passing once, and once only, through all the vertices. Furthermore, this path is unique.*

PROOF: There exists at least one vertex at which all the arcs are directed outwards, otherwise a path could be traced in the graph moving from vertex to vertex indefinitely, which is absurd since the graph has no circuits. Let a_1 be such a vertex; clearly a_1 is unique since another vertex a_1' with the same property must be joined, as the graph is complete, by an arc directed from a_1 to a_1' (say), and this is a contradiction.

Similarly, a unique vertex a_2 exists in the graph generated by $X - \{a_1\}$, at which all the arcs are directed outwards. Therefore, in the original graph, there exists an arc directed from a_1 to a_2.

Proceeding in this way, a unique path $a_1, a_2, a_3, \ldots$ can be constructed which passes once, and only once, through all the vertices of the graph.

PROPOSITION 1. *If the graph $(X, E \cup E^*)$ contains no circuits, there exists exactly one permutation f on X satisfying $E(f) = E$; if the graph contains a circuit, no permutation f exists with $E(f) = E$.*

PROOF: Obviously, if $(X, E \cup E^*)$ contains a circuit, then there is no permutation f satisfying $E(f) = E$.

If $(X, E \cup E^*)$ possesses no circuits, then, by the lemma, it contains a unique path $a_1, a_2, \ldots, a_n$ and, necessarily,

$$f = a_1 a_2 \cdots a_n.$$

Obviously, f is the unique permutation satisfying $E(f) = E$.   **Q.E.D.**

Write $f \geqslant g$ if $E(f) \supseteq E(g)$. Then:

PROPOSITION 2. *The relation $\geqslant$ is an order relation.*

PROOF: Obviously,

$$f \geqslant f,$$

$$\left. \begin{array}{c} f \geqslant g \\ g \geqslant f \end{array} \right\} \Rightarrow E(f) = E(g) \Rightarrow f = g,$$

$$\left. \begin{array}{c} f \geqslant g \\ g \geqslant h \end{array} \right\} \Rightarrow f \geqslant h.$$

Let (X, A) be a directed graph. *The transitive closure $\bar{A}$ of A in (X, A) is the set of all pairs xy of vertices which are joined by a path directed from x to y.*

PROPOSITION 3. *If f and g are permutations on a set X, there exists a unique permutation $f \vee g$ satisfying*

$$E(f \vee g) = \overline{E(f) \cup E(g)}.$$

$f \vee g$ is the least upper bound of f and g.

PROOF: By Proposition 1, it is sufficient to prove that the graph $(X, E \cup E^*)$, where $E = \overline{E(f) \cup E(g)}$, contains no circuits.

Suppose $(X, E \cup E^*)$ does contain a circuit. Then, since $(X, E \cup E^*)$ is both antisymmetric and complete, it contains a circuit of length 3, passing through the vertices a, b, and c (say).

We may assume $a < b < c$ otherwise relabel the vertices. If the circuit is a, b, c, then

$$ab, bc \in E; \qquad ca \in E^*,$$

which is impossible since the transitivity of E implies $ac \in E$.

If the circuit is, a, c, b, then

$$ac \in E; \qquad cb, ba \in E^*.$$

Therefore, bc, $ab \notin E(f) \cup E(g)$, and, hence,

$$cb, ba \in E^*(f) \cap E^*(g).$$

From transitivity, $ca \in E^*(f) \cap E^*(g)$, and, therefore,

$$ac \notin E(f) \cap E(g).$$

Relabeling f and g if necessary, this implies there exists a vertex x, $x \neq c$, such that

$$ax \in E(f), \qquad xc \in E(g), \qquad a < x < c.$$

However,

$$b < x \Rightarrow \begin{cases} b < x \\ bx \in E(f) \cup E^*(f) \end{cases} \Rightarrow bx \in E(f) \Rightarrow bc \in E: \qquad \text{contradiction};$$

$$b > x \Rightarrow \begin{cases} b > x \\ xb \in E(g) \cup E^*(g) \end{cases} \Rightarrow xb \in E(g) \Rightarrow ab \in E: \qquad \text{contradiction}.$$

Therefore, $f \vee g$ exists and, from the definition of closure, is the least upper bound of f and g.

PROPOSITION 4. *There exists a unique permutation $f \wedge g$, such that*

$$E^*(f \wedge g) = \overline{E^*(f) \cup E^*(g)}.$$

$f \wedge g$ is the greatest lower bound of f and g.

PROOF: By symmetry (inverting the order relation on X), this is an immediate consequence of Proposition 3.

Propositions 1, 2, and 3, therefore, prove that the permutohedron is a lattice.

5. DECOMPOSITION PROBLEMS

Consider a permutation f of the set $X = \{x_1, x_2, \ldots, x_n\}$.

The directed graph G_f is defined in the following way: the set of vertices is the set X and vertices x and y are joined by an arc (x, y), directed from x to y, if, and only if, $y = f(x)$; G_f is the union of disjoint *circuits*. On the other hand, let $T = \{t_1, t_2, \ldots, t_k\}$ be a set of transpositions on the set X. Then (X, T) is the undirected graph with vertex set X and edge set T.

EXAMPLE. $X = \{a, b, c, d\}$.

T consists of $t_1 = [ab]$, $t_2 = [bc]$, $t_3 = [bd]$.

$$f = t_1 t_2 t_3 = [ab][bc][bd] = \begin{pmatrix} a & b & c & d \\ b & d & a & c \end{pmatrix},$$

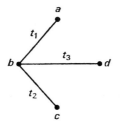
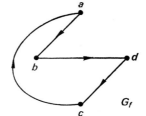

FIG. 5

$$g = t_2 t_3 t_1 = [bc][bd][ab] = \begin{pmatrix} a & b & c & d \\ d & a & b & c \end{pmatrix},$$

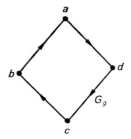

FIG. 6

$$gt_1 = \begin{pmatrix} a & b & c & d \\ d & a & b & c \end{pmatrix}[ab] = \begin{pmatrix} a & b & c & d \\ a & d & b & c \end{pmatrix},$$

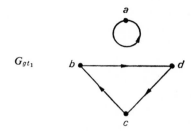

FIG. 7

$$t_1 g = [ab]\begin{pmatrix} a & b & c & d \\ d & a & b & c \end{pmatrix} = \begin{pmatrix} a & b & c & d \\ d & b & a & c \end{pmatrix}.$$

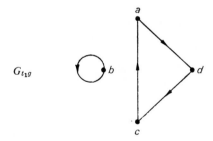

FIG. 8

THEOREM 1. *A set T of $n - 1$ transpositions generates the symmetric group S_n if, and only if, (X, T) is a tree.*

PROOF: (1) Suppose (X, T) is a tree. By Property 6 (Theorem 1, Chapter 3, Section 5), if a and b are distinct vertices of T, there exists a unique chain $[a\ x_1], [x_1\ x_2], \ldots, [x_k\ b]$ joining a to b. Since

$$[a\ b] = [x_k\ b][x_{k-1}\ x_k] \cdots [x_2\ x_3][x_1\ x_2][a\ x_1][x_1\ x_2]$$
$$\cdots [x_{k-1}\ x_k][x_k\ b],$$

every transposition $[a, b]$ is a product of elements of T. Hence, as S_n is generated by the transpositions $[a, b]$, $a, b \in X$, S_n is generated by the elements of T.

(2) If (X, T) is not a tree, by Property 3 (Theorem 1, Chapter 3, Section 5), it contains at least two nonempty edge-disjoint sets, X_1 and X_2 (say). Therefore, if $a \in X_1$, $b \in X_2$, the transposition $[a \ b]$ is not a product of elements of T.

LEMMA. *Let f be a permutation on X and $g = f \cdot [a \ b]$, where $[a \ b]$ is a transposition on X.*

The graph G_g can be obtained from the graph G_f by replacing the arcs $(a, f(a))$ and $(b, f(b))$ by $(a, f(b))$ and $(b, f(a))$, respectively.

(1) *If a and b are in different circuits in G_f, these circuits coalesce in G_g; moreover, if $\mu_f(x, y)$ is the set of vertices in the path with first vertex x and last vertex y ($\mu_f(x, y) = \varnothing$ if x and y are not connected),*

$$z \in \mu_f(x, y) \Rightarrow z \in \mu_g(x, y).$$

(2) *If a and b belong to the same circuit in G_f, then this circuit splits into two disjoint circuits in G_g; in addition,*

$$\left. \begin{array}{r} \mu_g(x, y) \neq \varnothing \\ z \in \mu_f(x, y) \end{array} \right\} \Rightarrow z \in \mu_g(x, y).$$

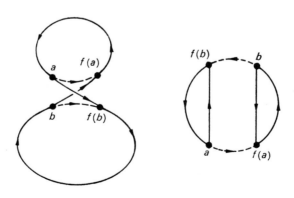

FIG. 9

PROOF: The lemma is obvious; if a and b belong to different circuits, then

$$g = f[a \quad b] = [a, fa, f^2a, \ldots][b, fb, f^2b, \ldots]h[ab]$$
$$= [a, fb, f^2b, \ldots, b, fa, f^2a, \ldots]h.$$

If a and b belong to the same circuit, then

$$g = f[a \quad b] = [a, fa, \ldots, b, fb, \ldots]h[a, b] = [a, fb, \ldots][b, fa, \ldots]h.$$

THEOREM 2 (Dénès). *Let $T = \{t_1, t_2, \ldots, t_{n-1}\}$ be a set of $n-1$ transpositions on a set X. The product $f = t_1 t_2 \cdots t_{n-1}$ is a circular permutation of length n if, and only if, (X, T) is a tree.*

PROOF:

(1) Suppose (X, T) is a tree. Consider the graphs G_g corresponding to the permutations

$$g_1 = t_1, \qquad g_2 = g_1 t_2, \qquad g_3 = g_2 t_3, \ldots, g_{n-1} = f.$$

By the lemma, the number of components of each of these graphs is

$$p(G_{g_1}) = n - 1,$$
$$p(G_{g_2}) = p(G_{g_1}) - 1 = n - 2,$$
$$p(G_{g_3}) = p(G_{g_2}) - 1 = n - 3,$$
$$\vdots$$
$$p(G_f) = p(G_{g_{n-2}}) - 1 = 1.$$

Hence, G_f is connected and is therefore a circuit, i.e., f is a circular permutation of length n.

(2) Suppose (X, T) is not a tree. By Property 3 (Theorem 1, Chapter 3, Section 5), (X, T) contains at least two nonempty edge-disjoint sets, X_1 and X_2 (say). Let $a \in X_1$ and $b \in X_2$. Then a and b do not belong to the same circuit in G_{g_1}; by the lemma, if a and b belong to different circuits in G_{g_i} then they belong to different circuits in $G_{g_{i+1}}$. Therefore, G_f is disconnected, and, hence, f is not a circular permutation of length n.

COROLLARY 5. *If f is a circular permutation of length n, the number $A(f)$ of ways of writing f as a product of $n-1$ transpositions is equal to n^{n-2}.*

PROOF: There are $(n - 1)!$ circular permutations of length n. Each of these permutations can be written in $A(f)$ ways, as a product of $n - 1$ transpositions. Therefore, the number of products of $n - 1$ transpositions, which are circular of length n, is $A(f)(n - 1)!$ On the other hand (Chapter 3, Section 5) there are n^{n-2} trees with n vertices. By the theorem, each tree can be associated with $n - 1!$ products of $n - 1$ transpositions, and each of these products is circular of length n. Hence, the number of products of $n - 1$ transpositions which are circular of length n is $(n - 1)!\, n^{n-2}$. Therefore,

$$A(f) = n^{n-2}. \quad \textbf{Q.E.D.}$$

We will now try to find the number of distinct circular permutations of length n, which, in the above sense, are associated with a tree (X, T).

Let $\bar{g} = t_1\, t_2\, \cdots\, t_{n-1}$ be a word on the $n - 1$ edges of the tree (X, T), and let $h_k(\bar{g})$ be the word on the edges, ordered as in $\bar{g}$, incident with the vertex k.

It is possible that, for distinct words, $\bar{g} = t_1\, t_2\, \cdots\, t_{n-1}$ and $\bar{g}' = t_1'\, t_2'\, \cdots\, t_{n-1}'$, $h_k(\bar{g}) = h_k(\bar{g}')$ for all k; for example, if, $X = \{x_1, x_2, x_3, x_4\}$, T consists of $t_1 = [x_1\ x_2]$, $t_2 = [x_2\ x_3]$, $t_3 = [x_3\ x_4]$, and $\bar{g} = t_2\, t_1\, t_3$, $\bar{g}' = t_2\, t_3\, t_1$, then

$$h_1(\bar{g}) = h_1(\bar{g}') = t_1, \qquad h_2(\bar{g}) = h_2(\bar{g}') = t_2\, t_1,$$
$$h_3(\bar{g}) = h_3(\bar{g}') = t_2\, t_3, \qquad h_4(\bar{g}) = h_4(\bar{g}') = t_3.$$

THEOREM 3 (Eden and Schützenberger). *Let* $\bar{f} = t_1\, t_2\, \cdots\, t_{n-1}$ *and* $\bar{g} = t_1'\, t_2'\, \cdots\, t_{n-1}'$ *be words on the* $n - 1$ *edges of the tree* (X, T). *Then* f *is equal to* g *if, and only if,*

$$h_k(\bar{f}) = h_k(\bar{g}) \qquad (k = 1, 2, \ldots, n).$$

PROOF:

(1)   We show, first of all, that if g is a permutation defined by

$$g = g_1\, t g_2\, t' g_3, \qquad t = [x_k\, a], \qquad t' = [x_k\, b],$$

then

$$x_k \in \mu_g(a, b).$$

By the lemma to Theorem 2, a and x_k belong to one circuit of $G_{g_1 t g_2}$ and b belongs to another. Therefore,

$$g_1 t g_2 t' = [a, f(a), \ldots, x_k, f(x_k) \cdots][b, f(b) \cdots]g'[x_k \, b]$$
$$= [x_k, f(b), \ldots, b, f(x_k) \cdots a, f(a) \cdots]g'.$$

Therefore,

$$x_k \in \mu_{g_1 t g_2 t'}(a, b),$$

and, again by the lemma,

$$x_k \in \mu_g(a, b).$$

(2)   Assume that

$$h_k(\bar{f}) \neq h_k(\bar{g}).$$

Then there exist transpositions $[x_k, a]$ and $[x_k, b]$, such that $[x_k, a]$ precedes $[x_k, b]$ in $\bar{f}$, and follows $[x_k, b]$ in $\bar{g}$. Therefore, from (1),

$$x_k \in \mu_f(a, b), \qquad x_k \in \mu_g(b, a)$$

and $f \neq g$.

(3) Assume that $h_i(\bar{f}) = h_i(\bar{g})$ for $i = 1, 2, \ldots, n$. If $\bar{g}'$ is obtained from $\bar{g}$ by interchanging consecutive, disjoint transpositions, then obviously $g' = g$. By this process, if possible, make the first term of $\bar{g}'$ the same as the first term of $\bar{f}$; then, if possible, make the second term of $\bar{g}'$ the same as the second term of $\bar{f}$, etc. Finally, either $\bar{f} = \bar{g}'$ and $f = g$ or

$$\bar{f} = \bar{f}_1[x_j \quad x_k]\bar{f}_2,$$
$$\bar{g}' = \bar{f}_1 \bar{g}_1[x_j \quad x_k]\bar{g}_2.$$

By the definition of $\bar{g}'$, the word $\bar{g}_1$ must contain a transposition which does not fix both x_j and x_k. Suppose $\bar{g}_1$ contains $[x_k, x_p]$. Then

$$\bar{f} = \bar{f}_1[x_j \quad x_k]\bar{f}_2'[x_k \quad x_p]\bar{f}_3',$$
$$\bar{g}' = \bar{f}_1\bar{g}_1'[x_k \quad x_p]\bar{g}_2'[x_j \quad x_k]\bar{g}_3'.$$

Therefore, $h_k(\bar{f}) \neq h_k(\bar{g})$, which is a contradiction. Therefore, $f = g$.

CONSEQUENCE 1. Let f be a circular permutation, and (X, T) a tree. How can we tell if f is a product

$$f = t_1 t_2 \cdots t_{n-1}, \qquad t_1, t_2, \ldots, t_{n-1} \in T,$$

of edges belonging to T?

For example, let

$$f = [1 \quad 3 \quad 8 \quad 2 \quad 5 \quad 4 \quad 6 \quad 7],$$

and suppose T consists of

$$t_1 = [x_1 \quad x_3], \qquad t_2 = [x_3 \quad x_8],$$

etc. (see Fig. 10).

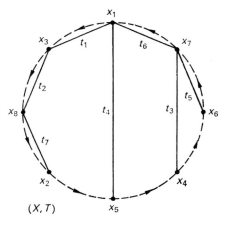

(X, T)

FIG. 10

Suppose f is a product of the $n-1$ edges belonging to (X, T). Then

$$h_1(\bar{f}) = [x_1 \quad x_7][x_1 \quad x_5][x_1 \quad x_3] = t_6 t_4 t_1.$$

The order of these factors is determined by (1) (Theorem 3, Section 5, Chapter 4), e.g., since

$$x_1 \in \mu_f(x_7, x_5),$$

$[x_1, x_7]$ precedes $[x_1, x_5]$ etc. Similarly,

$$h_2(\bar{f}) = t_7, \qquad h_3(\bar{f}) = t_1 t_2, \qquad h_4(\bar{f}) = t_3, \qquad h_5(\bar{f}) = t_4,$$

$$h_6(\bar{f}) = t_5, \qquad h_7(\bar{f}) = t_5 t_3 t_6, \qquad h_8(\bar{f}) = t_2 t_7.$$

$[h_k(\bar{f})$ can be quickly calculated from Fig. 8; starting at x_k, follow the circuit in a clockwise direction and write down the edges incident with x_k as they occur.] Since T is a tree and contains no cycles, there does exist a word $\bar{f}$ satisfying these equalities.

In this case,

$$\bar{f} = t_5\,t_3\,t_6\,t_4\,t_1\,t_2\,t_7\,.$$

Therefore,

$$f = [x_6 \quad x_7][x_7 \quad x_4][x_7 \quad x_1][x_1 \quad x_5][x_1 \quad x_3][x_3 \quad x_8][x_8 \quad x_2]$$

$$= \begin{pmatrix} 1 & 2 & 3 & 4 & 5 & 6 & 7 & 8 \\ 3 & 5 & 8 & 6 & 4 & 7 & 1 & 2 \end{pmatrix} = [1 \quad 3 \quad 8 \quad 2 \quad 5 \quad 4 \quad 6 \quad 7].$$

CONSEQUENCE 2. *Let (X, T) be a tree, and let $X = \{x_1, x_2, \ldots, x_n\}$. If $d(x_i)$ is the degree of x_i, $i = 1, 2, \ldots, n$, then the number of distinct circular permutations of length n, which, in the above sense, are associated with (X, T) is*

$$\prod_{i=1}^{n} d(x_i)!.$$

In particular, this result shows that a tree (X, T) cannot generate all the $(n-1)!$ circular permutations of length n.

REFERENCES

1. G. TH. GUILBAUD and P. ROSENSTIEHL, Analyse algebrique d'un scrutin, *Math. et Sci. Humaines* 4, 9–33 (1960).
2. P. ROSENSTIEHL, communication to NATO Colloquium on the theory of games, Toulon, 1966.
3. H. WIELANDT, "Finite Permutation Groups." Academic Press, New York, 1964.

CHAPTER 5

PÓLYA'S THEOREM

1. Counting Schemata Relative to a Group of Permutations of Objects

Let G be a group of permutations of the set $X = \{1, 2, \ldots, n\}$, and let φ be a mapping of X into

$$A = \{a_1, a_2, \ldots, a_m\}.$$

In this context, the elements of A are called *colors* and φ is said to be a *coloration*—each object i being "colored" with the color $\varphi(i)$.

Notice that if $g \in G$, the mapping φg is a coloration; for example,

$$\varphi g = \begin{pmatrix} 1 & 2 & 3 & 4 & 5 \\ b & c & a & a & c \end{pmatrix} \begin{pmatrix} 1 & 2 & 3 & 4 & 5 \\ 4 & 2 & 1 & 5 & 3 \end{pmatrix} = \begin{pmatrix} 1 & 2 & 3 & 4 & 5 \\ a & c & b & c & a \end{pmatrix}.$$

Two colorations φ_1 and φ_2 belong to the same *schema* if there exists $g \in G$ such that $\varphi_1 g = \varphi_2$. In this case, we write $\varphi_1 \sim \varphi_2$.

Clearly $\varphi_1 \sim \varphi_2$ is an equivalence relation:

$$\varphi \sim \varphi, \quad \text{since} \quad \varphi = \varphi e, \quad e \in G;$$

$$\varphi_1 \sim \varphi_2 \Rightarrow \varphi_2 \sim \varphi_1, \quad \text{since} \quad \varphi_1(g(x)) = \varphi_2(x) \Rightarrow \varphi_1(y) = \varphi_2(g^{-1}(y));$$

$$\varphi_1 \sim \varphi_2, \varphi_2 \sim \varphi_3 \Rightarrow \varphi_1 \sim \varphi_3,$$

$$\text{since} \quad \left. \begin{array}{l} \varphi_1(g(y)) = \varphi_2(y) \\ \varphi_2(h(x)) = \varphi_3(x) \end{array} \right\} \Rightarrow \varphi_1(gh(x)) = \varphi_3(x).$$

The problem we are concerned with here is that of counting the number of equivalence classes—or schemata—of this relation.

149

EXAMPLE 1: COLORATIONS OF THE CUBE. Let $X = \{1, 2, 3, 4, 5, 6\}$ be the set of faces of a cube, each of which is colored black or white. Thus $A = \{b, w\}$ (say). Two colored cubes are equivalent if one can be transformed into the other by a rotation. We shall now compute the number of schemata.

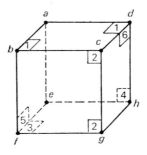

FIG. 1

In this example, a coloration of the cube is a mapping of X into A and G is the group of rotations of the cube:

about the axis	*abcd–efgh*:	[2 6 4 5]; [2 4][6 5]; [2 5 4 6]
about the axis	*bcfg–adhe*:	[1 5 3 6]; [1 3][5 6]; [1 6 3 5]
about the axis	*abfe–dcgh*:	[1 2 3 4]; [1 3][2 4]; [1 4 3 2]
about the axis	*a–g*:	[1 4 5][6 3 2]; [1 5 4][6 2 3]
about the axis	*b–h*:	[1 5 2][6 4 3]; [1 2 5][6 3 4]
about the axis	*c–e*:	[1 2 6][3 4 5]; [1 6 2][3 5 4]
about the axis	*d–f*:	[1 6 4][3 5 2]; [1 4 6][3 2 5]
about the axis	*ab–hg*:	[1 5][6 3][2 4]
about the axis	*bc–eh*:	[1 2][4 3][5 6]
about the axis	*cd–ef*:	[1 6][3 5][2 4]
about the axis	*ad–fg*:	[1 4][2 3][5 6]
about the axis	*bf–dh*:	[2 5][6 4][1 3]
about the axis	*cg–ae*:	[2 6][5 4][1 3]

Together with the identity [1][2][3][4][5][6], this makes 24 permutations in all.

There are 10 schemata:

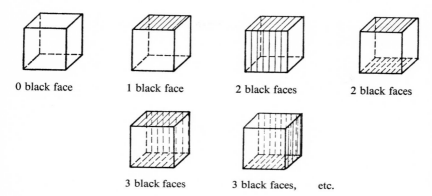

| 0 black face | 1 black face | 2 black faces | 2 black faces |

| | 3 black faces | 3 black faces,   etc. |

EXAMPLE 2: DISTRIBUTIONS. Let $X = \{1, 2, 3\}$, where 1 denotes a sphere S, and 2 and 3 denote indistinguishable cubes C. The objects 1, 2, and 3 are to be placed into distinct boxes a, b, and c.

A distribution is a mapping φ of X into $\{a, b, c\}$. The group G consists of $[1][2][3]$, $[1][2\ 3]$.

By inspection, there are 18 schemata:

$$
\begin{array}{ll}
SCC/\emptyset/\emptyset & C/SC/\emptyset \\
SC/C/\emptyset & C/S/C \\
SC/\emptyset/C & C/C/S \\
CC/S/\emptyset & C/\emptyset/SC \\
CC/\emptyset/S & \emptyset/SCC/\emptyset \\
S/CC/\emptyset & \emptyset/SC/C \\
S/C/C & \emptyset/CC/S \\
S/\emptyset/CC & \emptyset/S/CC \\
& \emptyset/C/SC \\
& \emptyset/\emptyset/SCC
\end{array}
$$

More generally, let $X = \{1, 2, \ldots, n\}$ be a set of objects and $A = (a_1, a_2, \ldots, a_n)$ a set of colors.

The objects are to be "colored" in all possible ways with the colors belonging to A. G is an arbitrary group of permutations on X. We now want to find (see Theorem 1, below) the number of schemata relative to G.

If $g \in G$, let $\lambda_i(g)$ denote the number of cycles of g of length i. Then the polynomial

$$P(G; x_1, x_2, \ldots, x_n) = \frac{1}{|G|} \sum_{g \in G} x_1^{\lambda_1(g)} x_2^{\lambda_2(g)} \cdots x_n^{\lambda_n(g)}$$

is called the *cycle index* of G.

THEOREM 1 (Pólya). *The number of schemata is*

$$P(G; m, m, \ldots, m).$$

EXAMPLE. In the case of the colorations of the cube,

$$P(G; x_1, x_2, x_3, x_4, x_5, x_6)$$
$$= \tfrac{1}{24}(x_1^{6} + 3x_1^{2}x_2^{2} + 6x_1^{2}x_4 + 6x_2^{3} + 8x_3^{2}).$$

By the lemma, therefore, the number of schemata is

$$P(G; 2, 2, 2, 2, 2, 2) = 10.$$

In the case of the distributions of a sphere and two cubes into three distinct boxes,

$$P(G ; x_1, x_2, x_3) = \tfrac{1}{2}(x_1^{3} + x_1 x_2).$$

By the lemma, therefore, the number of schemata is

$$P(G ; 3, 3, 3) = 18.$$

PROOF OF THEOREM: Consider a coloration $\varphi \in \Phi$, and a permutation $g \in S_n$. The mapping $\bar{g}$ of Φ into itself, defined by $\varphi \to \bar{g}(\varphi) = \varphi g$, is an injection since

$$\varphi \neq \varphi' \Rightarrow \varphi g \neq \varphi' g \Rightarrow \bar{g}(\varphi) \neq \bar{g}(\varphi').$$

Since Φ is finite, $\bar{g}$ is a bijection and $\bar{g}$ belongs to the set $\bar{S}$ of permutations of Φ.

The mapping $g \to \bar{g}$ of G into $\bar{S}$ is an injection, since

$$g \neq g' \Rightarrow g(k) \neq g'(k) \quad \text{for a } k \leqslant n,$$
$$\Rightarrow \varphi g \neq \varphi g' \quad \text{for some coloration } \varphi, \text{ in which } g(k) \text{ and}$$
$$g'(k) \text{ have different colors,}$$
$$\Rightarrow \bar{g} \neq \bar{g}'.$$

Therefore, $\bar{G} = \{\bar{g} \mid g \in G\}$ and G have the same cardinality. Furthermore, $\bar{G}$ is a subgroup of $\bar{S}$, since if $\bar{g}$, $\bar{h} \in \bar{G}$,

$$\bar{g} \cdot \bar{h}(\varphi) = \bar{g}(\varphi h) = \varphi(hg) = (\overline{hg})\varphi.$$

Hence, $\bar{g} \cdot \bar{h} \in \bar{G}$.

Colorations φ_1 and φ_2 are equivalent if $\varphi_1 = \bar{g}(\varphi_2)$, for some $\bar{g} \in \bar{G}$, that is, if they belong to the same orbit (Chapter 4, Section 3) of $\bar{G}$; the number of schemata is, therefore, equal to the number of orbits of $\bar{G}$ which is, by Theorem 2 (Chapter 4, Section 3),

$$|\mathscr{C}_G| = \frac{1}{|\bar{G}|} \sum_{\bar{g} \in \bar{G}} v(\bar{g}),$$

where $v(\bar{g})$ is the number of colorations φ satisfying $\varphi g = \varphi$, i.e., φ is *constant on each cycle of g*. Thus, $v(\bar{g})$ is equal to the number of mappings of the set of cycles of g (of cardinality $\lambda_1(g) + \lambda_2(g) + \cdots$) into the set of m colors. Hence,

$$|\bar{G}_G| = \frac{1}{|G|} \sum_{g \in G} m^{\lambda_1(g) + \lambda_2(g) + \cdots} = P(G; m, \ldots, m, m).$$

PÓLYA'S THEOREM. *The number of schemata with α_i objects colored a_i ($i = 1, 2, \ldots, m$) is*

$$\frac{1}{|G|} \sum_{\substack{\lambda_1, \lambda_2 \cdots \geqslant 0 \\ \lambda_1 + 2\lambda_2 + \cdots + n\lambda_n = n}} h_G(\lambda_1, \lambda_2, \ldots, \lambda_n) p_{\alpha_1 \alpha_2 \ldots}(\lambda_1, \lambda_2, \ldots, \lambda_n),$$

where $h_G(\lambda_1, \lambda_2, \ldots, \lambda_n)$ is the number of $g \in G$ of type $1^{\lambda_1} 2^{\lambda_2} \cdots n^{\lambda_n}$, and $p_{\alpha_1 \alpha_2 \ldots}(\lambda_1, \lambda_2, \ldots, \lambda_n)$ is the number of colorations, which are constant on the classes of a partition of X of type $1^{\lambda_1} 2^{\lambda_2} \cdots n^{\lambda_n}$, and in which exactly α_i objects are colored a_i ($i = 1, 2, \ldots, m$).

PROOF: Consider the set Φ_0 consisting of those colorations in which α_i objects are colored a_i ($i = 1, 2, \ldots, m$). The mapping $\varphi \to \bar{g}(\varphi) = \varphi g$, $\varphi \in \Phi_0$, of Φ_0 into itself is an injection, and therefore a permutation of Φ_0. As in the proof of Theorem 1, the number of schemata satisfying the conditions of the theorem is

$$|\mathscr{C}_G| = \frac{1}{|G|} \sum_{g \in G} v(\bar{g}),$$

where $v(\bar{g})$ is the number of colorations φ belonging to Φ_0, which are constant on each cycle of g. Therefore,

$$|\mathscr{C}_G| = \frac{1}{|G|} \sum_{\substack{\lambda_1, \lambda_2, \ldots, \lambda_n \geq 0 \\ \lambda_1 + 2\lambda_2 + \cdots = n}} \sum_{\substack{g \in G \\ g \text{ is of type} \\ 1^{\lambda_1} 2^{\lambda_2} \cdots}} p_{\alpha_1\alpha_2\ldots}(\lambda_1, \lambda_2, \ldots, \lambda_n).$$

Whence the formula.

EXAMPLE (Pólya). Given six similar spheres with

2   colored a,

1   colored b,

3   colored c,

how many ways are there of distributing them, one at each vertex, on an octahedron situated in three-dimensional space?

If the vertices of the octahedron are labeled 1, 2, 3, 4, 5, 6, the rotations of the octahedron induce permutations of the set $\{1, 2, \ldots, 6\}$; for example, a rotation of 90° about the diagonal 2–3 induces the permutation $g = [4\ 5\ 6][2][3]$:

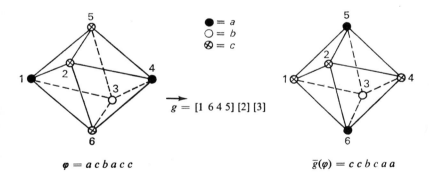

$$\varphi = a\,c\,b\,a\,c\,c \qquad\qquad \bar{g}(\varphi) = c\,c\,b\,c\,a\,a$$

Let G be the group of permutations induced by the rotations of the octahedron. The number of schemata is, by Pólya's theorem,

$$|\mathscr{C}_G| = \frac{1}{|G|} \sum_{\lambda_1 + 2\lambda_2 + \cdots + n\lambda_n = n} h_G(\lambda_1, \lambda_2, \ldots, \lambda_n)\, p_{\alpha_1\alpha_2\ldots}(\lambda_1, \lambda_2, \ldots, \lambda_n).$$

In the case of the octahedron we obtain

Types	Permutations of G	h_G	Acceptable colorations corresponding to the given type	p_α
1^6	[1] [2] [3] [4] [5] [6]	1	$(a)(a)(b)(c)(c)(c)$ $(a)(a)(c)(b)(c)(c)$ etc. ...	$\dfrac{6!}{2!1!3!} = 60$
$1^2 \cdot 4$	[4 5 1 6] [2] [3] [6 1 5 4] [2] [3] [1 2 4 3] [5] [6] [3 4 2 1] [5] [6] [2 5 3 6] [1] [2] [6 3 5 2] [1] [4]	6	$\varnothing$	0
$1^2 \cdot 2^2$	[4 1] [5 6] [2] [3] [1 4] [2 3] [5] [6] [2 3] [5 6] [1] [4]	3	$(b)(c)(a^2)(c^2)$ $(b)(c)(c^2)(a^2)$ $(c)(b)(a^2)(c^2)$ $(c)(b)(c^2)(a^2)$	4
2^3	[1 2] [3 4] [5 6] [1 3] [2 4] [5 6] [1 4] [2 5] [3 6] [1 4] [3 5] [2 6] [1 5] [4 6] [2 3] [1 6] [4 5] [2 3]	6	$\varnothing$	0
3^2	[1 2 5] [3 6 4] [1 2 6] [3 5 4] [1 3 5] [2 6 4] [1 3 6] [2 5 4] [1 5 2] [3 4 6] [1 6 2] [3 4 5] [1 5 3] [2 4 6] [1 6 3] [2 5 4]	8	$\varnothing$	0

Thus,

$$|\mathscr{C}_G| = \frac{1}{24}(60 + 3 \times 4) = 3.$$

2. COUNTING SCHEMATA RELATIVE TO AN ARBITRARY GROUP

Let $X = \{1, 2, \ldots, n\}$ be a set of *objects*; $A = \{a, b, c, \ldots\}$ a set of *m colors, and* φ a *coloration* of X by A. Then, if g is a permutation of X and h a permutation of A, $h\varphi g$ is again a coloration, for example,

$$h\varphi g = \begin{pmatrix} a & b & c & d \\ b & a & d & c \end{pmatrix} \begin{pmatrix} 1 & 2 & 3 & 4 & 5 \\ c & a & a & b & c \end{pmatrix} \begin{pmatrix} 1 & 2 & 3 & 4 & 5 \\ 4 & 2 & 1 & 5 & 3 \end{pmatrix}$$

$$= \begin{pmatrix} 1 & 2 & 3 & 4 & 5 \\ a & b & d & d & b \end{pmatrix}.$$

Let G be a group of permutations of $X \cup A$, fixing both X and A, i.e.,

$$\left. \begin{array}{l} (g, h) \in G \\ i \in X \end{array} \right\} \Rightarrow (g, h)i = g(i) \in X,$$

$$\left. \begin{array}{l} (g, h) \in G \\ a \in A \end{array} \right\} \Rightarrow (g, h)a = h(a) \in A.$$

We write

$$\{g/ (g, h) \in G\} = G_0, \qquad \text{group of permutations of } X,$$

$$\{h/ (g, h) \in G\} = H_0, \qquad \text{group of permutations of } A,$$

$$G \subseteq G_0 \times H_0.$$

Finally, let Φ_1 be a family of colorations, which is *closed* and *faithful* relative to G, i.e.,

$$(g, h) \in G, \qquad \varphi \in \Phi_1 \Rightarrow h\varphi g \in \Phi_1;$$

$$\left. \begin{array}{l} (g, h), (g', h') \in G \\ h\varphi g = h'\varphi g' \quad \text{for all} \quad \varphi \in \Phi_1 \end{array} \right\} \Rightarrow (g, h) = (g', h').$$

Two colorations φ_1 and φ_2 belonging to Φ_1 are *equivalent relative to G*,

$$\varphi_1 \sim \varphi_2 \quad (G),$$

if there exists a $(g, h) \in G$, such that $h\varphi_1 g = \varphi_2$.

Obviously, this is an equivalence relation, since

$$e\varphi e = \varphi,$$

$$h\varphi g = \varphi' \implies \varphi = h^{-1}\varphi'g^{-1},$$

$$\left.\begin{array}{c} h\varphi g = \varphi' \\ h'\varphi'g'' = \varphi'' \end{array}\right\} \implies \varphi'' = (h'h)\varphi(gg').$$

The equivalence classes of this relation are called *schemata relative to G*. The next theorem gives the number of schemata relative to G.

GENERAL THEOREM. *Let Φ_1 be a family of colorations, which is closed and faithful* relative to G. Then the number of schemata of Φ_1, relative to $G \subseteq S_X \times S_A$, is equal to*

$$\frac{1}{|G|} \sum_{(g,h)\in G} v(g,h),$$

where $v(g,h)$ is the number of $\varphi \in \Phi_1$ satisfying $\varphi = h\varphi g$, i.e., the elements in a cycle of g are colored with the colors (possibly repeated) occurring in just one cycle of h.

PROOF: If $(g,h) \in G$, let $t(g,h)$ be a mapping defined on Φ_1 as follows:

$$\varphi \to t(g,h)\varphi = h\varphi g.$$

Then,

$$\left.\begin{array}{c} \varphi, \varphi' \in \Phi_1 \\ \varphi(k) \neq \varphi'(k) \end{array}\right\} \implies h\varphi g(g^{-1}k) \neq h\varphi'g(g^{-1}k) \implies t(g,h)\varphi \neq t(g,h)\varphi'.$$

Thus, $t(g,h)$ is an injection and, hence, a permutation of Φ_1. Therefore, the number of schemata relative to G is equal to the number of orbits of $t(G)$. By Theorem 2 (Chapter 4, Section 1), and recalling that Φ_1 is faithful relative to G, this number is equal to

$$\frac{1}{|G|} \sum_{(g,h)\in G} \lambda_1\big(t(g,h)\big),$$

* Translator's note: The condition "that Φ_1 is faithful relative to G" is not necessary.

where $\lambda_1(t(g, h))$ is the number of $\varphi \in \Phi_1$ such that $h\varphi g = \varphi$; if i_1, i_2, i_3, ..., i_d is a cycle of g, then these elements are colored:

$$\varphi(i_1); \quad \varphi(i_2) = h^{-1}\varphi(i_1); \quad \varphi(i_3) = h^{-1}\varphi(i_2) = h^{-2}\varphi(i_1); \quad \text{etc.,}$$

and $h^{-d}(\varphi(i_1)) = \varphi(i_1)$. Hence $\lambda_1(t(g, h)) = v(g, h)$.

EXAMPLE 1. To obtain Pólya's theorem take

$G = G_0 \times \{e\}$,
$\Phi_1 = $ set of all mappings φ, such that the numbers $d_a = |\{i/\varphi(i) = a\}|$, $a \in A$ are preassigned.

To compute $R(1^{\lambda_1} 2^{\lambda_2} \cdots n^{\lambda_n}; 1^{\mu_1} 2^{\mu_2} \cdots m^{\mu_m})$ (Chapter 3, Section 4), take

$G = S_{X_1} \times S_{X_2} \times \cdots \times S_{A_1} \times S_{A_2} \times \cdots$,
$\Phi_1 = $ set of surjections of X into A.

EXAMPLE 2. Consider again the "*problème des ménages*" (Chapter 3, Section 3). $X = \{0, 1, 2, \ldots, n - 1\}$ is a set of husbands and $A = \{\underline{0}, \underline{1}, \underline{2}, \ldots, \underline{n-1}\}$ their respective wives; a bijection φ of X into A determines a *disposition* $\underline{0}$, $\varphi(0)$, $\underline{1}$, $\varphi(1)$, $\underline{2}$, ..., of husbands and wives around a circular table, etc.
Let g_p be the permutation of X defined by

$$g_p = \begin{pmatrix} 0 & 1 & 2 & \cdots & n-1 \\ 0+p & 1+p & 2+p & \cdots & n-1+p \end{pmatrix}.$$

[In this example, $i + p$ is understood to mean $i + p \pmod n$.]
If the label of each person i, $(\underline{i})$ is reduced to $i - p$ $(\underline{i - p})$, then the woman $\varphi(i) - p$ is seated to the right of the man $j = i - p$, and φ becomes φ', where

$$\varphi'(j) = \varphi(j + p) - p = g_p^{-1}\varphi g_p(j).$$

We count the number of dispositions of the husbands and wives around the table, where circular permutations of the labels of the husbands and wives are regarded as not producing essentially distinct seating plans; therefore, this number is equal to the number of schemata of Φ_1, relative to the group

$$G = \{(g_p, g_p^{-1})/p = 0, 1, 2, \ldots, n - 1\},$$

where Φ_1 is the set of bijections φ satisfying $\varphi(i) \neq i, i + 1$. Clearly, Φ_1 is closed and faithful relative to G. Therefore, the number of schemata is

$$\frac{1}{|G|} \sum_{g \in G} v(g_p, g_p^{-1}) = \frac{1}{n} \sum_{p=0}^{n-1} |\Phi_{1,p}|,$$

where $\Phi_{1,p}$ is the set of $\varphi \in \Phi_1$ such that, for each i, the images under φ of the elements of the cycle

$$i, i + p, i + 2p, \ldots$$

are colored, respectively,

$$\varphi(i), \varphi(i) + p, \varphi(i) + 2p, \ldots.$$

If $p = 0$ (Chapter 3, Section 3),

$$|\Phi_{1,0}| = |\Phi_1| = T(n) = \sum_{k=0}^{n} \frac{2n}{2n - k} (-1)^k \binom{2n - k}{k} (n - k)!.$$

Suppose $0 < p \leqslant n - 1$. Let $(n; p)$ denote the greatest common divisor of n and p, and write

$$q = \frac{n}{(n; p)}.$$

Then $pq \equiv 0 \pmod{n}$, and the cycles of g_p are all of length q. They are

$$C_0 = [0, p, 2p, \ldots, (q - 1)p],$$
$$C_1 = [1, 1 + p, 1 + 2p, \ldots, 1 + (q - 1)p],$$
$$\vdots$$
$$C_{s-1} = [s - 1, s - 1 + p, s - 1 + 2p, \ldots, s - 1 + (q - 1)p].$$

Since $sq = n$, the number of cycles of g_p is

$$s = \frac{n}{q} = \frac{n}{n/(n; p)} = (n; p).$$

If $\varphi \in \Phi_{1,p}$, the numbers $\varphi(0), \varphi(1), \ldots, \varphi(s - 1)$ completely define φ, since

$$\varphi(i + kp) = \varphi(i) + kp \qquad (i = 0, 1, \ldots, s - 1);$$

and furthermore, $\varphi(i) \neq i + \varepsilon, \varepsilon = 0, 1$, implies

$$\varphi(i + kp) = \varphi(i) + kp \neq (i + kp) + \varepsilon.$$

Thus, φ is completely determined by the sequence

$$\varphi(0), \varphi(1), \ldots, \varphi(s - 1)$$

($\varphi(i) \neq i, i + 1$), and conversely.

Now let π be a permutation of $\{0, 1, 2, \ldots, s - 1\}$, and let k be the cardinality of

$$\{i / i = 0, 1, 2, \ldots, s - 1; \quad \pi(i) = i \quad \text{or} \quad \pi(i) = i + 1\}.$$

For a given k, the number of permutations π is (Chapter 3, Section 3)

$$T^k(s) = \sum_{j=k}^{2s} (-1)^{j-k} \binom{j}{k} \frac{2s}{2s - j} \binom{2s - j}{j} (s - j)!.$$

If $\varphi(i) \in C_i$ (or C_{i+1}), there are $q - 1$ possible choices for $\varphi(i)$. If $\varphi(i) \in C_j$ ($j \neq i, i + 1$), there are q possible choices for $\varphi(i)$. Therefore, corresponding to each permutation π, the number of sequences $\varphi(0)$, $\varphi(1), \ldots$ satisfying $\varphi(i) \in C_{\pi(i)}$ and $\varphi(i) \neq i, i + 1$ is equal to

$$(q - 1)^k q^{s-k}.$$

Therefore,

$$|\Phi_{1, p}| = \sum_{k=0}^{s} T^k(s)(q - 1)^k q^{s-k}.$$

Finally, the number of schemata is

$$\frac{1}{n} T^0(n) + \frac{1}{n} \sum_{p=1}^{n-1} \sum_{k=0}^{(n; p)} T^k(n; p) \left(\frac{n}{(n; p)} - 1 \right)^k \left(\frac{n}{(n; p)} \right)^{(n; p) - k}$$

EXAMPLE 3. COUNTING KNOTS

(1) A knot (see Fig. 2) can be represented as a tangled loop of string (i.e., the string is knotted, and the two ends are spliced together) lying on a plane surface. It is an *alternating knot* if, when the string crosses itself, only two branches of the string intersect at the crossing and if, as one follows the string, it crosses itself alternately " over " and " under." In order to describe an alternating knot, choose a crossing

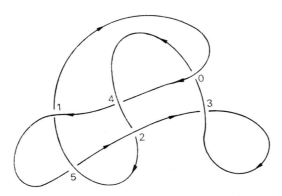

FIG. 2

and number it 0; next follow the string lying above this crossing in an arbitrary direction to the next crossing at which it lies above; number this new crossing 1, and, continuing in this way, number successively 2, 3, 4, ... the other crossings at which the string crosses over itself. The knot, so numbered, defines a bijection φ, in which the image $\varphi(i)$ of the upper crossing i is the undercrossing that immediately follows i. For example, in Fig. 2,

$$\varphi = \begin{pmatrix} 0 & 1 & 2 & 3 & 4 & 5 \\ \underline{4} & \underline{5} & \underline{3} & \underline{0} & \underline{2} & \underline{1} \end{pmatrix}.$$

(2) The knot has a "loop" if, following the string in the chosen direction, the upper crossing i is immediately followed by the lower crossing $\underline{i}$, i.e., $\varphi(i) = \underline{i}$. Similarly if $\varphi(i) = \underline{i+1}$, for example, $i = 2$ in Fig. 2. Such a loop can be removed by a simple twist of the string. Therefore, the alternating knots, considered in this example, are loopless.

In other words, the permutation φ defining the alternating knot satisfies the conditions of the "problème des ménages."

Moreover, if a different crossing was labeled 0, the knot would be defined by $\varphi' = g_p^{-1} \varphi g_p$, for some p; if, having labeled the crossing 0, the reverse of the chosen direction was taken, the knot would be defined by $\varphi' = h^{-1} \varphi h$, where

$$h = [n-1, 1][n-2, 2][n-3, 3] \cdots.$$

In each of these cases, the permutations φ and φ' are said to be *equivalent*. Therefore, an alternating knot with n crossings does not define just one permutation of degree n, but an equivalence class of permutations called the *signature* of the knot. The number of signatures of degree n is derived immediately from the formula in the last example. The following values have been obtained:

$n =$	3	4	5	6	7	8	9
	1	2	5	20	87	616	4843

(3)   A practical application is the following problem. We consider two alternating knots (I) and (II) of degree n; is it possible to transform one into the other by moving the string so that no new crossings are created, and all the crossings remain on the plane surface? For example, consider the knots

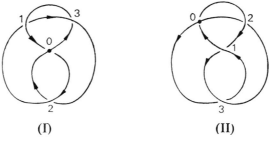

(I)   (II)

Fig. 3

It is easy to see that these knots are mirror images of one another, however, they both define the permutation

$$\varphi = \begin{pmatrix} 0 & 1 & 2 & 3 \\ 3 & 0 & 1 & 2 \end{pmatrix}.$$

Therefore, (I) can be transformed, with the above restrictions, into (II) (in other words, like Lewis Caroll, one might say, the knot has passed through the looking glass).

I : $\varphi = [03]\,[142]$

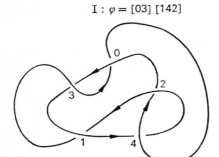

II : $\varphi = [03142]$

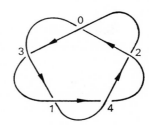

III : $\varphi = [02143]$

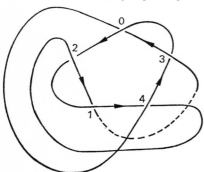

IV : $\varphi = [02413]$

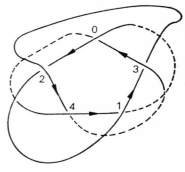

V : $\varphi = [04321]$

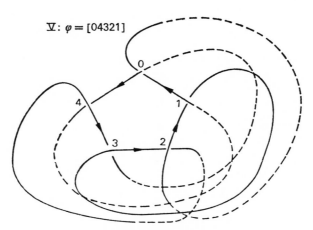

FIG. 4. The 5 signatures of degree 5.

163

(4)   More generally, a *knot is of order n* if there exists a representation of the knot on a plane surface containing n crossings, and no other such representation exists containing less than n crossings.

Knots N and N' are *identical* if there exists a homeomorphism of R^3 into itself which maps N onto N'.

There exist numerous works on the classification of knots [see, for example, the bibliography of Fox (1963) or, more recently, Neurwirth's book, "Knot Groups," Princeton University Press, 1965]. Unfortunately, not every knot is alternating. Moreover, not every signature of degree n defines an alternating knot of order n (see Fig. 4).

3. A THEOREM OF DE BRUIJN

Let $\varphi_0 \in \Phi$ be a coloration, and suppose φ_0 belongs to the schemata $\bar{\varphi}_0$ of Φ, relative to the group G of permutations of X; then $\bar{\varphi}_0$ is, by definition, the set

$$\{\varphi \mid \varphi \in \Phi, \varphi g = \varphi_0 \qquad \text{for some} \quad g \in G\}.$$

If $\bar{\varphi}_0$ remains fixed, relative to a permutation h of the colors, we write $h\varphi_0 \in \varphi_0 \, G$.

Suppose each object colored a is assigned a weight $w(a) > 0$; if, in a coloration $\varphi \in \bar{\varphi}$, r_i objects are colored a_i, the *weight* of φ is defined by

$$W(\varphi) = w(a_1)^{r_1} w(a_2)^{r_2} \cdots w(a_m)^{r_m}.$$

Let $W(\bar{\varphi})$ denote the common value of $W(\varphi)$, $\varphi \in \bar{\varphi}$. In the next theorem we calculate the sum of the weights of those schemata fixed by h.

THEOREM (de Bruijn).   *The sum of the weights of the schemata, which are invariant with respect to a permutation h of the colors, is*

$$\sum W(\bar{\varphi}) = P(G \, ; p_1, p_2, \ldots, p_n),$$

where

$$P(G_i; x_1, x_2, \ldots, x_n) = \frac{1}{|G|} \sum_{g \in G} (x_1)^{\lambda_1} (x_2)^{\lambda_2} \cdots (x_n)^{\lambda_n}$$

is the cycle index of G, and

$$p_k = \sum_{\substack{a \in A \\ h^k a = a}} w(a) w(ha) w(h^2 a) \cdots w(h^{k-1} a).$$

PROOF:

(1) Let $\bar{\varphi}_0$ be a schema that is invariant with respect to h, i.e., $h\varphi_0 \in \varphi_0 G$.

Then

$$\bar{\varphi} = \bar{\varphi}_0 \Rightarrow \varphi \in \varphi_0 G \Rightarrow h\varphi = h\varphi_0 g = \varphi_0 g_0 g \in \varphi_0 G \Rightarrow \overline{h\varphi} = \bar{\varphi}_0.$$

Conversely,

$$\overline{h\varphi} = \bar{\varphi}_0 \Rightarrow \begin{cases} h\varphi \in \varphi_0 G \\ h\varphi_0 \in \varphi_0 G \end{cases} \Rightarrow \begin{cases} \varphi = h^{-1}\varphi_0 g \\ \varphi_0 = h^{-1}\varphi_0 g_0 \end{cases} \Rightarrow \varphi = h^{-1}\varphi_0 g_0(g_0^{-1}g) \in \varphi_0 G$$

$$\Rightarrow \bar{\varphi} = \bar{\varphi}_0.$$

Therefore, $\bar{\varphi}_0$ consists of the colorations φ satisfying $\overline{h\varphi} = \bar{\varphi}_0$.

(2) The mapping $\varphi \to h\varphi g = \bar{g}(\varphi)$ is a permutation of Φ, since

$$\varphi \neq \varphi' \Rightarrow h\varphi g \neq h\varphi' g;$$

the orbit of the group $\bar{G} = \{\bar{g} / g \in G\}$, containing φ_0, is

$$O(\varphi_0) = \{\varphi / h\varphi g = \varphi_0 \quad \text{for some} \quad g \in G\} = \{\varphi / \overline{h\varphi} = \bar{\varphi}_0\}$$

Therefore, by Theorem 1 (Chapter 4, Section 3)

$$\sum W(\bar{\varphi}) = \sum_{\varphi / h\varphi \in \varphi G} \frac{W(\varphi)}{|O(\varphi)|} = \frac{1}{|\bar{G}|} \sum_{\varphi / h\varphi \in \varphi G} W(\varphi)|\bar{G}_\varphi|$$

$$= \frac{1}{|G|} \sum_{g \in G} \sum_{\substack{\varphi / h\varphi \in \varphi G \\ h\varphi g = \varphi}} W(\varphi) = \frac{1}{|G|} \sum_{g \in G} \sum_{\substack{\varphi \\ h\varphi g = \varphi}} W(\varphi)$$

(since $h\varphi g = \varphi$ implies $h\varphi = \varphi g^{-1} \in \varphi G$).

(3) Let $(Y_1, Y_2, \ldots, Y_s)$ be the partition of X determined by the cycles of a permutation g of X of type $1^{\lambda_1} 2^{\lambda_2} \cdots n^{\lambda_n}$. Choose a sequence $y_1, y_2, \ldots, y_s$, such that

$$y_i \in Y_i \qquad (i = 1, 2, \ldots, s).$$

If $h\varphi g = \varphi$, and if $k(i)$ is the length of the cycle of g containing y_i, the elements of Y_i are colored

$$\varphi(y_i) = a, \qquad \varphi g^{-1}(y_i) = ha, \ldots, \varphi g^{-k(i)}(y_i) = h^k a = a.$$

Therefore, assuming

$$p_k(a) = w(a)w(ha)w(h^2 a) \cdots w(h^{k-1}a) \qquad \text{if} \quad a = h^k a,$$
$$= 0 \qquad \text{if} \quad a \neq h^k a_i;$$

$$W(\varphi) = \prod_{i=1}^{s} p_{k(i)}(\varphi(y_i)).$$

(4)   Let

$$A_k = \{a/\ a \in A;\ h^k a = a\} = \{a_k^{\ 1}, a_k^{\ 2}, \ldots, a_k^{p_k}\}.$$

Then, if

$$\varphi(y_i) = a_{k(i)}^{j_i} \qquad (i = 1, 2, \ldots, s),$$

φ is completely determined by the s-tuple $(j_1, j_2, \ldots, j_s)$ and

$$\sum_{h\varphi g = \varphi} W(\varphi) = \sum_{(j_1, j_2, \ldots, j_s)} p_{k(1)}(a_{k(1)}^{j_1}) p_{k(2)}(a_{k(2)}^{j_2}) \cdots p_{k(s)}(a_{k(s)}^{j_s})$$

$$= \prod_{i=1}^{s} [p_{k(i)}(a_{k(i)}^1) + p_{k(i)}(a_{k(i)}^2) + \cdots]$$

$$= \prod_{i=1}^{s} \sum_{a \in A_{k(i)}} w(a)w(ha) \cdots w(h^{k(i)-1}a)$$

$$= \prod_{i=1}^{s} p_{k(i)} = (p_1)^{\lambda_1}(p_2)^{\lambda_2} \cdots (p_n)^{\lambda_n}.$$

The theorem follows immediately.

EXAMPLE. The group G of rotations of the cube is given in Example 1 (Chapter 5, Section 1). By inspection,

$$P(G; x_1, x_2, x_3, x_4, x_5, x_6)$$
$$= \tfrac{1}{24}(x_1^{\ 6} + 3x_1^{\ 2}x_2^{\ 2} + 6x_1^{\ 2}x_4 + 6x_2^{\ 3} + 8x_3^{\ 2}).$$

The number of ways of coloring the cube, using two colors a and b, so that the coloration is invariant relative to an interchange of the colors, is obtained by taking $h = (a, b)$, $w(a) = w(b) = 1$, viz.,

$$P(G\ ; 0, 2, 0, 2, 0, 2) = \tfrac{1}{24}(6 \cdot 2^3) = 2.$$

We now want to find the number of ways of coloring the cube with the colors a_1, a_1', a_2, a_2', a_3, a_3', so that the coloration is invariant relative to $h = [a_1\ a_1'][a_2\ a_2'][a_3\ a_3']$, i.e., h has the same effect as one of the rotations of the cube. For simplicity, write $w(a_i) = w_i$, $w(a_i') = w_i'$; then

$$p_1 = 0,$$

$$p_2 = 2w_1w_1' + 2w_2w_2' + 2w_3w_3',$$

$$p_3 = 0,$$

$$p_4 = 2(w_1)^2(w_1')^2 + 2(w_2)^2(w_2')^2 + 2(w_3)^2(w_3')^2,$$

$$p_5 = 0,$$

$$p_6 = 2(w_1)^3(w_1')^3 + 2(w_2)^3(w_2')^3 + 2(w_3)^3(w_3')^3.$$

The required number is the coefficient of $w_1w_1'w_2w_2'w_3w_3'$ in the polynomial

$$P(G; p_1, p_2, p_3, p_4, p_5, p_6) = \frac{1}{24}\, 6\, p_2^3$$

$$= \frac{2^3}{4}\, (w_1w_1' + w_2w_2' + w_3w_3')^3$$

$$= 2 \sum_{\alpha+\beta+\gamma=3} \frac{3!}{\alpha!\beta!\gamma!}\, (w_1w_1')^\alpha(w_2w_2')^\beta(w_3w_3')^\gamma.$$

Therefore, there are 12 acceptable colorations.

COROLLARY 1 (Pólya). *Suppose $w(a_i) = w_i$. Then the sum of the weights of the schemata, relative to a group G of permutations of the objects, is*

$$\sum W(\bar\varphi) = P\left(G; \sum_{i=1}^m w_i, \sum_{i=1}^m w_i^2, \sum_{i=1}^m w_i^3 \cdots\right).$$

The coefficient of $w_1^{\alpha_1} w_2^{\alpha_2} \cdots w_m^{\alpha_m}$ in this polynomial, is equal to the number of schemata $\bar\varphi$, such that each $\varphi \in \bar\varphi$ has α_i objects colored a_i $(i = 1, 2, \ldots, m)$.

PROOF: In the theorem, take h to be the identity permutation.

COROLLARY 2 (de Bruijn). *Let G_0, H_0 be permutation groups on X, A, respectively. As above, assume weights are assigned to each schema relative to the group*

$$G = G_0 \times H_0.$$

Then the sum of the weights of the schemata is

$$\sum W(\bar{\varphi}) = \frac{1}{|H_0|} \sum_{h \in H_0} P(G_0; p_1(h), p_2(h), \ldots),$$

where

$$p_k(h) = \sum_{h^k a = a} w(a)w(ha) \cdots w(h^{k-1}a).$$

PROOF: By Theorem 1 (Chapter 4, Section 3),

$$\sum W(\bar{\varphi}) = \frac{1}{|H_0|} \sum_{h \in H_0} \sum_{\substack{\bar{\bar{\varphi}} \\ h\varphi \in \varphi G}} W(\bar{\varphi})$$

where $\bar{\varphi}$ denotes a schema relative to G_0, which is invariant with respect to h.

APPLICATION: COUNTING UNLABELED GRAPHS.   Let G be a group of permutations of a set X and let $\mathscr{P}_2'(X)$ be the set of unordered pairs $\{x, y\}$, $x, y \in X$, $x \neq y$.

A permutation $g \in G$ induces a mapping $\bar{g}$ of $\mathscr{P}_2'(X)$ into $\mathscr{P}_2'(X)$, defined by

$$\bar{g}(x, y) = \{g(x), g(y)\}.$$

$\bar{g}$ is clearly a permutation of $\mathscr{P}_2'(X)$, and the set of these permutations $\bar{g}$ forms a group, denoted by $G^{(2)}$. Two graphs (X, U) and (X, V) are *isomorphic* if there exists a permutation g of the vertices of one of them satisfying

$$(x, y) \in U \Leftrightarrow \bar{g}(x, y) \in V.$$

An *unlabeled graph* is a class of isomorphic graphs. We shall now calculate the number of nonisomorphic graphs with n vertices.

$\mathscr{P}_2'(X)$ is the set of $n(n-1)/2$ edges of the complete graph (without loops or multiple edges) on the vertices $x_1, x_2, \ldots, x_n$. Each mapping

$n =$	Unlabeled graphs	$P(S_n^{(2)}; x_1, x_2, \ldots)$	$P(S_n^{(2)}; 2, 2, \ldots)$	$P(S_n^{(2)}; 0, 2, 0, 2, \ldots)$	$P(S_n^{(2)}; 1 + w, 1 + w^2, \ldots)$
2		x_1	2	0	$1 + w$
3		$\frac{1}{6}(x_1^3 + 3 x_1 x_2 + 2 x_3)$	4	0	$1 + w + w^2 + w^3$
4		$\frac{1}{24}(x_1^6 + 9 x_1^2 x_2^2 + 8 x_3^2 + 6 x_2 x_4)$	11	1	$1 + w + 2 w^2 + 3 w^3 + 2 w^4 + w^5 + w^6$

169

φ, of $\mathscr{P}_2'(X)$ into a set $A = \{a, b\}$ of two colors, determines a graph (X, U) defined by

$$U = \{u/\, u \in \mathscr{P}_2'(X),\ \varphi(u) = a\}.$$

Clearly the correspondence $\varphi \rightarrow (X, U)$ is a bijection.

Therefore, the number of unlabeled graphs with n vertices is

$$P(S_n^{(2)}; 2, 2, 2, \ldots) \qquad \text{(Theorem 1, Section 1).}$$

The number of unlabeled graphs, which are isomorphic to their complements, is

$$P(S_n^{(2)}; 0, 2, 0, 2, \ldots) \qquad \text{(Theorem, Section 3).}$$

The number of decompositions of the edges of the complete graph with n vertices is

$$\tfrac{1}{2}P(S_n^{(2)}; 2, 2, 2, \ldots) + \tfrac{1}{2}P(S_n^{(2)}; 0, 2, 0, 2, \ldots) \qquad \text{(Corollary 2, Section 3).}$$

The number of unlabeled graphs with n vertices and m edges is equal to the coefficient of w^m in

$$P(S_n^{(2)}; 1 + w, 1 + w^2, 1 + w^3, \ldots) \qquad \text{(Corollary 1, Section 3).}$$

The number of self-complementary graphs with m edges and n vertices is equal to the coefficient of w^m in

$$P(S_n^{(2)}; 0, 2w, 0, 2w^2, 0, 2w^3, \ldots) \qquad \text{(Theorem, Section 3).}$$

The corresponding directed graphs are counted by replacing $S_n^{(2)}$ by the cartesian product $S_n \otimes S_n$.

4. Computing the Cycle Index

We now describe [9], without detailed proofs, the polynomial $P(G) = P(G; x_1, x_2, \ldots, x_n)$ for certain special groups G.

(1) The *symmetric group* S_n is of order $n!$

From Cauchy's formula (Chapter 4, Section 2),

$$P(S_n) = \sum_{\lambda_1 + 2\lambda_2 + \cdots = n} \frac{1}{\lambda_1!\, 2^{\lambda_2}\lambda_2! \cdots n^{\lambda_n}\lambda_n!}\, x_1^{\lambda_1} x_2^{\lambda_2} \cdots x_n^{\lambda_n}.$$

(2)   The *alternating group* A_n consists of the even permutations of n objects and is of order $\frac{1}{2}n!$. Clearly,

$$P(A_n) = \sum_{\lambda_1 + 2\lambda_2 + \cdots = n} \frac{1 + (-1)^{\lambda_2 + \lambda_4 + \lambda_6 \cdots}}{\lambda_1! 2^{\lambda_2} \lambda_2! \cdots n^{\lambda_n} \lambda_n!} x_1^{\lambda_1} x_2^{\lambda_2} \cdots x_n^{\lambda_n}.$$

(3)   The *identity group* $E_n = \{e\}$, consisting of just the identity permutation e, has cycle index

$$P(E_n) = x_1{}^n.$$

(4)   The *cyclic group* C_n of order n is generated by the circular permutation $[1 \ 2 \ \cdots \ n]$ and

$$P(C_n) = \frac{1}{n} \sum_{k/n} \varphi(k)(x_k)^{n/k},$$

where $\varphi(k)$ (Chapter 3, Section 3) is the Euler function.

(5)   The *dihedral group* D_n, of order $2n$, is generated by the two permutations $[1, 2, \ldots, n]$ and $[1, n][2, n-1] \cdots$.
If $n = 2p$, then

$$P(D_{2p}) = \frac{1}{4p} \sum_{k/2p} \varphi(k)(x_k)^{(2p/k)} + \frac{1}{4}(x_2{}^p + x_1{}^2 x_2^{p-1}).$$

If $n = 2p + 1$, then

$$P(D_{2p+1}) = \frac{1}{2(2p+1)} \sum_{k/2p+1} \varphi(k) x_k^{(2p+1)/k} + \frac{1}{2} x_1 x_2{}^p.$$

(6)   The *direct product* $G \times H$. Let G and H be permutation groups of order r and s, respectively. Suppose the elements of G act on a set X of cardinality p, and the elements of H act on a set Y of cardinality q, which is disjoint from X. Then the direct product $G \times H$ acts on $X \cup Y$ and is defined by

$$(g, h)z = \begin{cases} gz & \text{if} \quad z \in X, \\ hz & \text{if} \quad z \in Y. \end{cases}$$

The order of $G \times H$ is clearly rs, and

$$P(G \times H) = P(G) \times P(H).$$

(7)   The *Cartesian product* $G \otimes H$. $G \otimes H$ acts on the set $X \times Y$ and is defined by

$$(g, h)(x, y) = (gx, hy).$$

Its order is rs.

Suppose x belongs to a cycle C_g of g of length k, and y belongs to a cycle C_h of h of length l; then (x, y) belongs to a cycle of (g, h) of length $m(k, l)$, where $m(k, l)$ is the least common multiple of k and l; altogether, there are $(k; l)$ cycles of (g, h), corresponding to C_g and C_h, where $(k; l)$ is the greatest common divisor of k and l. Hence,

$$P(G \otimes H) = \frac{1}{|G| \cdot |H|} \sum_{g,h} \prod_{k,l} x_{m(k, l)}^{(k; l)\lambda_k(g)\lambda_l(h)}.$$

(8)   The *wreath product* $G[H]$ acts on the set $X \times Y$, and consists of the permutations $t(g; h_1, h_2, \ldots, h_p)$ defined by

$$t(g; h_1, h_2, \ldots, h_p)(x_i, y) = (gx_i, h_i y).$$

Its order is $r \cdot s^p$. Pólya proved

$$P(G[H]) = P(G; p_1(H), p_2(H), \ldots),$$

where $p_k(H) = P(H; x_k, x_{2k}, x_{3k}, \ldots)$.

(9)   The *exponentiation* H^G acts on the set Y^X of mappings of X into Y, and consists of the permutations $t(g, h)$ defined by

$$t(g, h)f(x) = hfg(x).$$

Its order is rs and

$$P(H^G) = \frac{1}{|G| \cdot |H|} \sum_{\substack{g \in G \\ h \in H}}^{q^p} \prod_k x_k^{\lambda_k(g, h)},$$

where

$$\lambda_1(g, h) = \prod_k \sum_{r/k} [r\lambda_r(h)]^{\lambda_k(g)},$$

and

$$\lambda_k(g, h) = \frac{1}{k} \sum_{r/k} \mu(r, k)\lambda_1(g^r, h^r).$$

References

1. N. G. DE BRUIJN, Color patterns that are invariant under a given permutation of the colors, *J. Comb. Theory* **2**, 418–421 (1967).
2. N. G. DE BRUIJN, Generalization of Pólya's fundamental theorem in enumerative combinatorial analysis, *Indagationes Mathematicae* **21**, 59–69 (1959).
3. N. G. DE BRUIJN, Pólya's theory of counting, *in* "Applied Combinatorial Mathematics" (E. F. Beckenbach, ed.), Chapter 5. Wiley, New York, 1964.
4. G. DEMOUCRON, Y. MALGRANGE, and R. PERTUISET, Reconnaisance et construction de représentations planaires topologiques, *Rev. Française Informat. Recherche Operationnelle* **30**, 33–47 (1964).
5. J. DÉNÈS, The representation of a permutation as the product of a minimal number of transpositions, and its connection with the theory of graphs, *Math. Inst. Hungarian Acad. Sci.* **4**, 63–70 (1959).
6. M. EDEN and M. P. SCHÜTZENBERGER, Remark on a theorem of Dénès, *Math. Inst. Hungarian Acad. Sci. Ser. A* **7**, 353–355 (1962).
7. E. N. GILBERT, Knots and classes of ménage permutations, *Scripta Math.* **22**, 228–233 (1956).
8. F. HARARY, The number of linear, directed, rooted and connected graphs. *Trans. Amer. Math. Soc.* **78**, 445–463 (1955).
9. F. HARARY, Graphical enumeration problems, *in* "Applied Combinatorial Mathematics" (E. F. Beckenbach, ed.). Wiley, New York, 1966.
10. F. HARARY, Unsolved problems in the enumeration of graphs, *Math. Inst. Hungarian Acad. Sci.* **5**, 63–95 (1960).
11. I. KAPLANSKY and J. RIORDAN, Le problème des ménages, *Scripta Math.* **12**, 113–124 (1946).
12. A. LEMPEL, S. EVEN, and I. CEDERBAUM, An algorithm for planarity testing of graphs, Séminaire de Rome, "Theory of Graphs," pp. 215–232. Dunod, Paris, 1958.
13. G. PÓLYA, Kombinatorische Anzahlbestimmungen für Gruppen, Graphen, und chemische Verbindungen, *Acta Math.* **68**, 145–254 (1937).
14. J. RIGUET, Notice sur quelques principes fondamentaux d'énumération, *in* "Theorie des Graphes" (C. Berge, ed.). Dunod, Paris, 1958.
15. P. G. TAIT, "Scientific Papers," Vol. 1, p. 287. Cambridge Univ. Press, London, 1898.

INDEX

Mathematics in Science and Engineering

A Series of Monographs and Textbooks

Edited by RICHARD BELLMAN, *University of Southern California*

1. T. Y. Thomas. Concepts from Tensor Analysis and Differential Geometry. Second Edition. 1965

2. T. Y. Thomas. Plastic Flow and Fracture in Solids. 1961

3. R. Aris. The Optimal Design of Chemical Reactors: A Study in Dynamic Programming. 1961

4. J. LaSalle and S. Lefschetz. Stability by by Liapunov's Direct Method with Applications. 1961

5. G. Leitmann (ed.). Optimization Techniques: With Applications to Aerospace Systems. 1962

6. R. Bellman and K. L. Cooke. Differential-Difference Equations. 1963

7. F. A. Haight. Mathematical Theories of Traffic Flow. 1963

8. F. V. Atkinson. Discrete and Continuous Boundary Problems. 1964

9. A. Jeffrey and T. Taniuti. Non-Linear Wave Propagation: With Applications to Physics and Magnetohydrodynamics. 1964

10. J. T. Tou. Optimum Design of Digital Control Systems. 1963.

11. H. Flanders. Differential Forms: With Applications to the Physical Sciences. 1963

12. S. M. Roberts. Dynamic Programming in Chemical Engineering and Process Control. 1964

13. S. Lefschetz. Stability of Nonlinear Control Systems. 1965

14. D. N. Chorafas. Systems and Simulation. 1965

15. A. A. Pervozvanskii. Random Processes in Nonlinear Control Systems. 1965

16. M. C. Pease, III. Methods of Matrix Algebra. 1965

17. V. E. Benes. Mathematical Theory of Connecting Networks and Telephone Traffic. 1965

18. W. F. Ames. Nonlinear Partial Differential Equations in Engineering. 1965

19. J. Aczel. Lectures on Functional Equations and Their Applications. 1966

20. R. E. Murphy. Adaptive Processes in Economic Systems. 1965

21. S. E. Dreyfus. Dynamic Programming and the Calculus of Variations. 1965

22. A. A. Fel'dbaum. Optimal Control Systems. 1965

23. A. Halanay. Differential Equations: Stability, Oscillations, Time Lags. 1966

24. M. N. Oguztoreli. Time-Lag Control Systems. 1966

25. D. Sworder. Optimal Adaptive Control Systems. 1966

26. M. Ash. Optimal Shutdown Control of Nuclear Reactors. 1966

27. D. N. Chorafas. Control System Functions and Programming Approaches (In Two Volumes). 1966

28. N. P. Erugin. Linear Systems of Ordinary Differential Equations. 1966

29. S. Marcus. Algebraic Linguistics; Analytical Models. 1967

30. A. M. Liapunov. Stability of Motion. 1966

31. G. Leitmann (ed.). Topics in Optimization. 1967

32. M. Aoki. Optimization of Stochastic Systems. 1967

33. H. J. Kushner. Stochastic Stability and control. 1967

34. M. Urabe. Nonlinear Autonomous Oscillations. 1967

35. F. Calogero. Variable Phase Approach to Potential Scattering. 1967

36. A. Kaufmann. Graphs, Dynamic Programming, and Finite Games. 1967

37. A. Kaufmann and R. Cruon. Dynamic Programming: Sequential Scientific Management. 1967

38. J. H. Ahlberg, E. N. Nilson, and J. L. Walsh. The Theory of Splines and Their Applications. 1967

39. Y. Sawaragi, Y. Sunahara, and T. Nakamizo. Statistical Decision Theory in Adaptive Control Systems. 1967

40. R. Bellman. Introduction to the Mathematical Theory of Control Processes Volume I. 1967 (Volumes II and III in preparation)

41. E. S. Lee. Quasilinearization and Invariant Imbedding. 1968

42. W. Ames. Nonlinear Ordinary Differential Equations in Transport Processes. 1968

43. W. Miller, Jr. Lie Theory and Special Functions. 1968

44. P. B. Bailey, L. F. Shampine, and P. E. Waltman. Nonlinear Two Point Boundary Value Problems. 1968

45. Iu. P. Petrov. Variational Methods in Optimum Control Theory. 1968

46. O. A. Ladyzhenskaya and N. N. Ural'tseva. Linear and Quasilinear Elliptic Equations. 1968

47. A. Kaufmann and R. Faure. Introduction to Operations Research. 1968

48. C. A. Swanson. Comparison and Oscillation Theory of Linear Differential Equations. 1968

49. R. Hermann. Differential Geometry and the Calculus of Variations. 1968

50. N. K. Jaiswal. Priority Queues. 1968

51. H. Nikaido. Convex Structures and Economic Theory. 1968

52. K. S. Fu. Sequential Methods in Pattern Recognition and Machine Learning. 1968

53. Y. L. Luke. The Special Functions and Their Approximations (In Two Volumes). 1969

54. R. P. Gilbert. Function Theoretic Methods in Partial Differential Equations. 1969

55. V. Lakshmikantham and S. Leela. Differential and Integral Inequalities (In Two Volumes). 1969

56. S. H. Hermes and J. P. LaSalle. Functional Analysis and Time Optimal Control. 1969

57. M. Iri. Network Flow, Transportation, and Scheduling: Theory and Algorithms. 1969

58. A. Blaquiere, F. Gerard, and G. Leitmann. Quantitative and Qualitative Games. 1969

59. P. L. Falb and J. L. de Jong. Successive Approximation Methods in Control and Oscillation Theory. 1969

60. G. Rosen. Formulations of Classical and Quantum Dynamical Theory. 1969

61. R. Bellman. Methods of Nonlinear Analysis, Volume I. 1970

62. R. Bellman, K. L. Cooke, and J. A. Lockett. Algorithms, Graphs, and Computers. 1970

63. E. J. Beltrami. An Algorithmic Approach to Nonlinear Analysis and Optimization. 1970

64. A. H. Jazwinski. Stochastic Processes and Filtering Theory. 1970

65. P. Dyer and S. R. McReynolds. The Computation and Theory of Optimal Control. 1970

66. J. M. Mendel and K. S. Fu (eds.). Adaptive, Learning, and Pattern Recognition Systems: Theory and Applications. 1970

67. C. Derman. Finite State Markovian Decision Processes. 1970

68. M. Mesarovic, D. Macko, and Y. Takahara. Theory of Hierarchial Multilevel Systems. 1970

69. H. H. Happ. Diakoptics and Networks. 1971

70. Karl Astrom. Introduction to Stochastic Control Theory. 1970

71. G. A. Baker, Jr. and J. L. Gammel (eds.). The Padé Approximant in Theoretical Physics. 1970

72. C. Berge. Principles of Combinatorics. 1971

73. Ya. Z. Tsypkin. Adaptation and Learning in Automatic Systems. 1971

74. Leon Lapidus and John H. Seinfeld. Numerical Solution of Ordinary Differential Equations. 1971

75. L. Mirsky. Transversal Theory, 1971

In preparation

Harold Greenberg. Integer Programming

E. Polak. Computational Methods in Optimization: A Unified Approach

Thomas G. Windeknecht. A Mathematical Introduction to General Dynamical Processes

Andrew P. Sage and James L. Melsa. System Identification

R. Boudarel, J. Delmas, and P. Guichet. Dynamic Programming and Its Application to Optimal Control

William Stenger and Alexander Weinstein. Methods of Intermediate Problems for Eigenvalues Theory and Ramifications